the
green
travel guide

231 164

the green travel guide

Paul Jenner and Christine Smith

Editor: Roni Jay

crimson

This edition first published in Great Britain 2008 by

Crimson Publishing, a division of Crimson Business Ltd
Westminster House
Kew Road
Richmond
Surrey
TW9 2ND

10 9 8 7 6 5 4 3 2 1

13-digit ISBN 978 1 905410 31 6

British Library Cataloguing in Publication Data

A CIP record for this book can be obtained from the British Library.

Printed and bound by Mega Printing, Turkey

The paper used for the text pages of this book is FSC certified
FSC (The Forest Stewardship Council) is an international
network to promote responsible management of the world's forests

This book is dedicated to all the wonderful people who struggle to protect the world's wild animals and wild places.

Contents

PART 1

How to be a green traveller

Welcome to *The Green Travel Guide*, the book for all who love travel and want to continue enjoying it without damaging the very things they've come to experience. Here in *Part One* we take a look at the theory of what it means to be a green traveller. We give you the principles that will allow you to carry on travelling without harming the environment. In fact, done in the right way, your trips can actually improve the environment. We tell you how. Later, in *Part Two,* we'll give you full details of some actual green holidays.

Ten ways to reduce your travel impact on the environment

- Always choose the nearest suitable destination. In other words, for a beach holiday opt for Britain or France or Spain rather than the Caribbean.
- Always choose coach or rail in preference to flying or driving.
- When you do drive it should be in a car capable of 50 miles to the gallon or better.
- If your car has empty seats, share – or switch to coach/rail.
- Cross the Channel by the Channel Tunnel.
- Don't cruise, except on sailing boats.
- Always offset your pollution generously.
- Choose accommodation that minimises the impact on the environment.
- Avoid environmentally damaging holiday activities – quad bikes, jet skis and hunting, for example.
- Don't make business trips if the telephone, internet and video conferencing can do the job just as well.

Introduction

Is there such a thing as a green holiday?

You're probably planning to take a foreign holiday this year. Just as you always do. And you're probably planning a few short breaks on top. And why not? You work hard and your holidays are part of your reward as well as a much-needed opportunity to relax and recharge your batteries.

But maybe your usual excited anticipation has been spoiled by a new concern. Given what we now know about the environmental impact, is it really acceptable to travel far? Or even at all?

There's the waste of energy for a start. A two-week long-distance holiday can burn as much power as you normally do in the other 50 weeks of the year. Then there's the pollution. Not to mention the environmental impact of all the hotel development. And the cultural impact of thousands of westerners arriving in traditional communities. And so it goes on.

So can there actually be such a thing as green travel? And if so, what is it?

What is green travel?

As far as we're concerned, green travel, at its palest, is simply travel that doesn't do too much harm to the environment. In other words, you don't have to be a do-gooder, if you haven't got the inclination or the time. Don't worry. You can still enjoy yourself. Just don't be a do-harmer.

But there's also a deeper shade of green. In other words, travel that's positively *good* for the environment.

Just lying on a beach, for example, *can* be pale green. It all depends how you got to the beach. If it's a beach in Britain, not too far from where you live and you got there by train then that's pale green. You're not harming the environment. Even if the beach is in Spain then, provided you travelled by train or coach or fuel-efficient car with all seats taken, that's still got a green tinge to it. But if you've chosen the Caribbean to do your sunbathing that's very definitely not green.

What kinds of travel could be categorised as positively beneficial? We'll be describing some of those holidays in detail later in the book. The kinds of things we have in mind are trips that give value to wilderness and wild animals and put money into the pockets of local people as an incentive to preserve those things. But be careful. Flying to, say, South America to plant trees may not be as green as it appears. Unless you stay a long time and plant an awful lot of trees you're just not going to justify the eight or so tonnes of pollutants produced by your round trip.

But we're not going to get too hair shirt here. There's no particular reason to single out travel as being more dangerous for the environment than many other human activities. Just to give you one example, the clearing of forests is estimated to contribute almost a fifth of the most dangerous pollutants being put into the atmosphere. That's far more than all the holidays taken in one year all over the world.

We're in *favour* of travel. In the right circumstances it can actually benefit the planet. It can help save those forests. If we're going to be realistic we all have to accept that *everything* human beings do has some sort of impact on the environment. That's unavoidable. Come to that, elephants can have a pretty enormous impact on the environment, too. Yes, travel causes environmental problems. But so does eating. What we have to do is tackle *every* aspect of environmental damage.

So we're certainly not saying you can't go on holiday unless you're doing something good for the planet. We'd simply say – at the very least, try to minimise the damage and pay to offset the rest. In *Part One* of the book you'll find out how. In *Part Two* we describe some specific holidays.

And those ones *are* pretty deep green. But they're hugely enjoyable as well.

So *how* exactly can travel be good for the environment? Well, let's look at some of the things that really *are* destroying the planet.

POVERTY

Here's a statistic that may astonish you. Every year, one million tonnes of bush meat are taken from Central Africa. That's estimated to be six times the maximum sustainable yield. In Sarawak the level of hunting is thought to be 13 times the maximum sustainable yield. Why is it happening? In a word, poverty. For many people, bush meat is the cheapest available protein. In fact, for those with the necessary skills, it's virtually free. You're never going to convince them to stop hunting if the alternative is starvation. They need an income to buy food and where are they going to get it? *Tourism*, that's where. When we pay to visit wild places and to see wild animals, and when that money goes to local people, so those things have value in their eyes. And they will preserve them. Out of self-interest.

Other environmentally damaging consequences of poverty include the use of poisons and explosives to fish coral reefs, the felling of trees for firewood, and 'slash and burn' farming methods – that's to say, clearing forest land by fire, farming it for a short while until the soil is degraded, then moving to a new patch. Tourism can help prevent those, too.

IGNORANCE

We all went to school, we can all read and write, and we can all get the latest news. We know about things like pollution, habitat loss and wildlife extinctions. But education costs money. In many areas of the world that are now vital to the survival of the planet as we know it, people don't have the benefit of education. They can't read, they can't write. They find it difficult to know what's going on. Where is that money for education going to come from? There are several possibilities but *tourism* is very definitely one of them. It's the world's biggest industry and it generates enormous revenues.

INDUSTRIAL POLLUTION

People have to make a living somehow. If they don't make it from tourism they'll make it by some other means including logging, mining, farming, manufacturing and so on. Of course, we need all those activities to take place somewhere in the world. But when it comes to sensitive areas, when it comes to wilderness and wildlife habitat, which is the least worst? We think *ecotourism*, properly managed, is a better option than clearing tropical forest for, say, palm oil plantations or factories.

HUMAN RIGHTS ABUSES

Regrettably, in much of the world, people's natural rights are denied them. But, at least, when countries open up to *tourism* it becomes just that little bit harder. In a closed country anything can happen. When tourists are around, when they're in touch with local people, tyrants have to be a little bit more careful. What's more, where there's no free press, contact with outsiders is one of the ways of learning the truth about the world.

THE BUILT ENVIRONMENT

Buildings are part of the environment, too. Think of Venice, the Colosseum, the Parthenon, the Pyramids, the Taj Mahal, Machu Pichu and all the other man-made wonders of the world. Money is needed to preserve those buildings and it's *tourism* that generates it.

The rock and the hard place

When you make a trip, whether for a holiday, business or to visit friends and family, you're contributing to the world's number one industry. If travel, especially air travel, was severely curtailed, as some environmentalists would prefer, the effect on the global economy would be catastrophic. We just can't ignore the positive contribution tourism has made to living standards in so many parts of the world. Including our own.

Of course, travel and tourism can also be destructive. Extremely destructive. Nobody disputes that. In fact, that's the whole point of this book. In the well-known phrase, we're between a rock and a hard place. We're going to show you the green way through the space.

Chapter 1

Green ways of getting there

FACTS YOU NEED TO KNOW

- Transport accounts for about a fifth of global energy consumption.
- Transport accounts for about a quarter of certain key pollutants put into the atmosphere.
- Leisure travel accounts for around half of those key pollutants.
- On a fortnight's holiday to South Africa you'd burn roughly as much energy as you typically would for all your travelling during the other 50 weeks of the year.
- Coaches, trains and fuel-efficient cars with all seats taken use significantly less energy than planes or boats.

So you're going on a green holiday. Maybe it's a photographic safari in one of Africa's great national parks, maybe you're watching orcas from a sea kayak, maybe you're rescuing wolves. But how are you going to get there? The problem is this. The greater the distance and the further you go the higher the energy consumption and the greater the pollution.

Let's have a look at some of the possibilities.

On foot

Walking would be about the most environmental way of going on holiday. But is it really practical? Well, it could be. You can probably get along at about three miles an hour. Which means that if you're fit you could hike as much as 20 miles a day. It doesn't sound very much but it means that over a two-week holiday you could cover around 250 miles.

The most environmental of all would be to follow a circular route, beginning and ending at your house. Alternatively, you could walk 250 miles in a straightish sort of line and then return by coach or train. We've got suggestions for specific walking holidays in *Part Two*, but to give you an idea, you could walk from London to the New Forest (100 miles) in a week, spend a further week there, and return by coach. From Birmingham you could make a similar trek to the Black Mountains. And from Manchester you could make a marvellous tour of the Peak District.

On yer bike

The idea of cycling *on* holiday is growing in popularity. In fact, about 450,000 Britons do that each year and some 16% of adults have taken a cycling holiday at some time. But what about cycling *to* your holiday? Well, why not? If you enjoy cycling you may as well go the whole hog.

How long would it take to get somewhere nice where people wear berets and eat Camembert? To give you an idea, in the Tour de France they think nothing of 100-mile stages (and they still get most of the day off). A few people have even cycled that distance in under an hour. Mind you, they did cheat a bit by having someone drive in front to cut the wind resistance. Not actually allowed under the rules of the Tour de France and nothing very green about it, either! Realistically, if you cycle a bit but aren't an athlete, and you're loaded up with panniers, you should think in terms of a more relaxing 10 miles an hour and 50 to 60 miles a day.

Okay, that's not very much compared with a car but it's still a significant distance. As a Londoner, you could easily be in Brighton the same day. London to Dover (for the ferry), Manchester to Snowdonia, and

Newcastle upon Tyne to the Lake District are all possible in two days.

It's really a question of attitude. Instead of thinking in terms of getting to your destination as fast as possible, you make the journey an enjoyable part of your whole holiday experience.

There is a compromise. Use your bike for touring around your chosen holiday area but get it there by coach, train, ferry or, at a pinch (and provided every seat is taken) car. Theoretically, you can also take your bike on a plane but that would completely wipe out any green benefits, so we're not including that option here. (However, if you do decide to fly your bike, the consensus is to put it in a hard, not soft, travel bag, otherwise, it's likely to arrive damaged.)

Are walking and cycling pollution free?

Some people have suggested that walking and cycling might not be as green as they first appear, since both require additional amounts of energy in the form of food. In fact, brisk walking burns about 350 more calories per hour than just sitting in a car (although your metabolic rate would go up in a tense driving situation). And cycling uses about 500 calories more.

According to the World Watch Institute, a cyclist burns around 35kCal to travel one mile while a car burns roughly 50 times more or 1,800kCal.

How does that translate into carbon dioxide? A typical family saloon produces around 1,000lb of carbon dioxide in covering 1,000 miles compared with 32lb for a cyclist eating raw food and about 37lb on cooked food (that is, allowing for the fuel to make a hot meal). With the average 1.56 people in the car, the bicycle is 17 times more efficient. And, in fact, the World Watch figure seems to be based on a speed of about 20mph, which most people aren't capable of. At a more leisurely 10mph the carbon dioxide production of the cyclist would be halved.

BIKES BY COACH

The greenest method of transporting your bike would be by coach. But how exactly could you get your bike onto a coach? European Bike Express has the solution. It's a coach service operating within the UK and to France, Spain and Italy, using trailers specially equipped for bicycles.

There are four routes. Within the UK the service runs from Thornaby via various large towns to Folkestone. Then from Calais there's a service to Agen (France), another to Ampuriabrava (Spain – Costa Brava) and lastly to Cavallino (Italy). You can return from your drop-off point (usually 17 days later) or you can cycle to a different pick-up.

The coaches have fully reclining seats, plenty of leg room, air conditioning, WC, hot food and even a lounge area.

The cost is around £200 return per passenger and bike, with little difference between short and long journeys (because when someone gets off there's usually no one else to get on).

➜ www.bike-express.co.uk
➜ Tel. 01430 422 111

BIKES ON TRAINS

Now that Britain's railways are run by various different companies there's no uniform policy on bicycles. Except that under the National Conditions of Carriage operators aren't obliged to carry them. You'll have to contact each company individually [*see the section on trains below*]. But, generally, there's not too much problem. As regards the Channel Tunnel, you and your bicycle will be loaded onto a special minibus and thus onto Eurostar.

BIKES ON FERRIES

Again, each ferry company has its own policy and price structure but you shouldn't encounter any problems.

Yes, but how do you get your suitcase on a bicycle?

Although cycling is often associated with camping, you'll be much better off if you stay in hotels of some sort. Because the truthful answer is that you can't carry very much. A good pair of rear panniers, for example, will only carry around 40 litres. (For some reason, panniers are measured like bottles of milk which doesn't mean a great deal to the average person – but you get the rough idea.) You can also have front panniers but you wouldn't want them to be very big because it starts to become hard to steer. Say around 15 litres the pair. Your maps and documents can go in a handlebar bag and, if you're really loaded, you can sling something across the top of the rear panniers and even wear a small backpack for good measure. Even all that will only add up to what you could get in an average suitcase. Not a lot of room for a tent, sleeping bag, stove and all the rest of the camping gear. So to keep the whole thing fun we recommend you find a nice little hotel at the end of each day with a hot bath, a comfortable bed and a decent meal.

Sur votre vélib, monsieur!

Paris has got the right idea. It's installed some 10,000 bicycles in 750 docking stations (with more on the way). The idea is that you rent your *vélib* (from *vélo* meaning bicycle and *liberté* meaning freedom), cycle where you want and leave it there, in a different docking station. Within a short while of introduction, four million people had used *vélibs*, clocking up about 100,000 journeys a day. Hopefully, by the time you read this, it'll be a lot more.

The amazing Mr Jason Lewis

If you don't think human-powered transport could be a practical proposition, take courage – and inspiration – from Jason Lewis. In October 2007 he arrived back in London on his pedal-boat having gone all around the world via (as far as possible) the equator. Which goes to prove that non-polluting long-distance journeys *are* possible.

Of course, there is the little matter that it took him 13 years. But knock off the 10 or so months he spent recovering from being run over, as well as the time he spent working to raise funds, and maybe you could do it a little quicker.

But for the moment the record is his, by reason of being the only person ever to have done it. The means of transport he used were bike, in-line skates, kayak and pedal boat. When those seemed too complicated he simply swam or walked.

The whole journey amounted to 46,505 miles (74,842km), an average of around 3,500 miles (5,635km) a year. It was an amazing feat.

Jason was 26 when he set out from London with his companion Steve Smith, the man who actually came up with the idea in the first place. But Steve quit in Hawaii in 1999.

The Pacific was subdued by a 26ft wooden pedalo. "Pedalling across an ocean is such a huge expanse of space and time," he told a journalist from the *Independent*. "Sometimes you can get a little dispirited. But you have to forget about getting to the other end and concentrate on the here and now."

Excellent advice for any green traveller!

Coach

Coach travel isn't normally thought of as being very cool. Especially not if you have to sit up all night.

But things are changing. There *are* luxurious coaches and, what's more, they're about the greenest form of motorised transport that exists. For every 100kg of carbon dioxide pollution per passenger from an aircraft, a train would emit 25, a reasonably fuel-efficient car (with its average 1.56 passengers) 20, and a coach around 17. Of course, it all depends on the precise specification, load factors, the length and nature of the journey and how direct a route it's possible to follow. But that's a ball park sort of figure.

But if there's little difference between a coach and a fuel efficient car with two passengers, there's nevertheless another reason to favour the coach. Consider this. The M25 is 118 miles long and varies between three and four lanes each side. If cars are travelling at 30 mph the M25 can accommodate about 33,000 of them, containing (typically) 53,000 passengers. At speeds of 50mph there's only room for half the number of cars (because the stopping distance between them increases) and at 70mph the number drops to not much more than a quarter. But a coach can do much better. If there were nothing but coaches on the M25 its capacity would be increased by more than 1,000 per cent. Yes, that's not a misprint. Without cutting down another tree, destroying another acre of farmland or steam-rollering another particle of asphalt, the capacity of the M25 – and roads generally – could be increased ten times over.

So a coach should be one of your first thoughts. Apart from relative greenness, coaches do have their own special advantages. For a start, there's a good chance of being collected and dropped off at a more convenient place than a train can manage. And, of course, if you've opted for an organised coach tour you don't have to worry about transfers – a significant consideration if you've got a lot of luggage and it's pouring with rain, to boot. Then there's the fact that you've always got something to look at out of the window. From an aircraft, clouds are fascinating and beautiful but they can get monotonous.

So how far can you go before you start banging your head on the seat in front and wishing you'd taken the plane after all? Unless you're a businessman in a hurry, there's almost nowhere in the UK that can't reasonably be reached by coach. Coach is also a practical proposition for most journeys in Europe. Let's say you want to go to the French Alps for a spot of cross-country skiing. You could get a coach leaving London at tea-time and be in your resort at lunchtime the following day. That's not bad.

Europe's largest coach network is Eurolines. It actually comprises 32 independent coach companies all collaborating to cover some 500 destinations as far away as Moscow and Morocco.

To find out how to get around on public transport in the UK take a look at: www.traveline.org.uk or telephone 0871 200 22 33 (charged at 10p per minute from a BT landline).

➜ For an itinerary on train, coach (or, indeed, car) see www.transportdirect.info

➜ For National Express see www.nationalexpress.com Tel. 08705 80 80 80

→ For Eurolines see www.nationalexpress.com/eurolines
Tel. 08705 80 80 80

→ For Stagecoach see www.nationalexpress.com or telephone Traveline (see above)

The world's longest regular coach service

Step aboard OzBus in London and you can step off 12 weeks later close to Sydney Opera House. You'll have travelled about 15,000 miles (24,000km) half-way round the world and without flying – other than the short hop from Bali to Darwin (while the coach makes its way on a cargo boat). And this isn't a one-off organised for gap-year students. This is a regular service (at least, that's the plan).

So let's see how it stacks up against the usual way of getting to Australia. Well, flying is certainly more direct. You'd do about 10,000 miles (16,000km) and be responsible for about five-and-a-half tonnes of carbon dioxide one way. By comparison, if you went by OzBus, even though the overland route is 50 per cent longer, you'd still get to Australia on about two tonnes. So that's a useful saving.

And it would be a real adventure. In fact, for your two tonnes you'd not only see Sydney but some fantastic places in between – Prague, Transylvania, Istanbul, Bam (Iran), the Taj Mahal, Everest base camp (you have to get out of the coach for that), Sumatra and Uluru (Ayers Rock), to name just a few.

The price is a little bit more than the typical flight – £3,750. But when you consider that it's spread over 84 nights and includes food and accommodation in tents and budget hotels (with a few nights sitting up in the coach) it's pretty good value at around £45 a day.

Website: www.oz-bus.com
Tel. 0208 641 1443

Several OzBus clients have published blogs. See http://ozbusdiaries.blogspot.com or just put OzBus into your search engine and see what you can find.

Motorbikes

Ah yes, the open road. Everything you need packed into a couple of panniers. A throaty roar and you're off on your energy-saving tour of Europe. And, surely, you can permit yourself a smug little smile for such a green mode of transport? Well, no, actually you can't. Astonishingly, motorbikes aren't, broadly speaking, greener than cars. As we'll see in a moment, there are cars with fuel consumption figures of the order of 60mpg or better and carbon dioxide emissions of not much more than 100g/km. By comparison, most motorbikes give between 40 and 50mpg and around 150g/km of carbon dioxide. A few are better and quite a lot are worse. What's more, motorbikes carry only one or two people as against four or more for a car. So when it comes to per capita emissions, the fully loaded car wins hands down. Motorbike manufacturers should be ashamed of themselves.

Cars

Can you take a green or, at least, greenish holiday by car? It sounds impossible, given that road transport accounts for a fifth of all the UK's carbon emissions. And every right-thinking green seems implacably opposed to the car. Nevertheless, the answer is that – walking and cycling aside – the car can be as green as the best alternatives or even better. It all depends on two things. The type of car. And the number of people in it.

If you're going to travel as a family of, say, four in a fuel-efficient car then you'll actually cause less pollution damage to the environment than if you travel by either train or coach and far, far less than by flying.

Environmentalists tend to quote figures based on the least efficient cars and the UK average of 1.56 passengers. But there's no reason you can't do much better than that. (And if you haven't got a family of, say, four, why not take a couple of friends with you?)

If you make a 1,000km journey by train, for example, you'll be responsible for around 160kg of carbon. If you take an inter-city train you'll be responsible for about 110kg. And if you take an express coach you'll be responsible for around 80kg of carbon. But if you have one of the most efficient cars – say, the Toyota Prius – and if you go on holiday as a family of four, your 1,000km journey will work out at around 35kg of carbon each. (Note that, in calculating the figure, we've assumed you'll get only 75 per cent of the petrol consumption figure on the rolling road because that tends to be the reality.)

Of course, figures vary and there are plenty of other factors to be taken into account, like the relative damage done by building the trains, coaches, cars, railway lines and roads in the first place. But the message is clear. It isn't necessarily more virtuous to travel by coach or train than to travel by car.

It certainly is more virtuous, however, to travel by train or coach than as the only person in a gas-guzzler. The Range Rover V8, for example, does only 19 miles to the gallon and produces 352 grams of carbon dioxide per kilometre on the rolling road. In reality, if you drove it on a 1,000km holiday and were the only person in the car you could be responsible for anything up to 500kg of carbon dioxide. That's 14 times more than the family of four in the Prius.

What can you do?

The most important thing you can do when changing your car is to get a model with a low fuel consumption and, therefore, a low carbon footprint. As we've seen, the difference between the highest and lowest consumers is enormous. Aim for under 120g/km of carbon dioxide. At the moment there are around 10 petrol and 25 diesel models to choose from, which is plenty. And the manufacturers will make more if they see there's a demand. So create a demand. Help save the planet and buy a low-carbon car. Have nothing whatsoever to do with fast acceleration, high top speeds or Jeremy Clarkson.

The tragedy is that cars today are, generally, no more fuel-efficient than cars a quarter of a century ago. The manufacturers say they're trying their best but it just isn't true. They're still concentrating on performance, not fuel efficiency. We can change that if we use our chequebooks the right way.

In particular, take a look at hybrids such as the Toyota Prius. Essentially, the Prius uses petrol on the open road but its electric motor for short city hops and the two together for rapid acceleration. The batteries get recharged when you brake and when the petrol engine is operating.

Make and model	Fuel consumption (mpg)*	CO_2 (g/km)*
Petrol cars producing 120g/km CO_2 or less		
Toyota Prius	65.7	104
Honda Civic Hybrid	61.4	109
Citroen C1	61.4	109
Toyota Aygo	61.4	109
Peugeot 107	61.3	109
Smart Fortwo 50	60.1	113
Daihatsu Charade	58.9	114
Vauxhall Corsa	58.8	115
Smart Roadster	57.6	116
Daihatsu Sirion	56.5	118
Smart Fortwo 61	56.5	120
Diesel cars producing 120 g/km CO_2 or less		
Citroen C1	68.9	109
Toyota Aygo	68.9	109
Peugeot 206	65.6	112
Citroen C2	65.7	113
Fiat Panda	56.7	114
Renault Clio	65.7	115
Vauxhall Corsa	65.6	115
Citroen C3	64.2	115
Ford Fiesta 1560	64.2	116
Mini Cooper D	64.2	118
Hyundai Getz	62.8	118
Vauxhall Corsa	64.2	119
Fiat Gd Punto	62.8	119
Ford Fiesta 1399	62.8	119
Ford Fusion 1399	62.8	119

Ford Fusion 1560	62.8	119
Toyota Yaris	62.8	119
Volkswagen Polo	62.8	119
Citroen C3	62.8	120
Citroen C4	62.8	120
Renault Modus	62.8	120
Renault Megane	62.8	120
Peugeot 207 1398	62.7	120
Peugeot 207 1560	62.7	120
Skoda Fabia	61.4	120

*Note: These are official test figures achieved on a rolling road. In the real world of day-to-day driving you might only achieve around three-quarters of the claimed fuel consumption.

You can check the CO_2 emissions of every car on this website: www.vcacarfueldata.org.uk

Apart from anything else, you'll save a lot of money on fuel. Let's say you're taking a family holiday in the south of France. By road, allowing for sightseeing, that's a round trip of something like 2,000 miles (3,200km). Do it in a Prius and you'll use only half as much petrol as a similar but conventional saloon, a saving of something like £120.

PETROL OR DIESEL?

Theoretically, diesel engines should be far more efficient than petrol engines and, therefore, greener. Unfortunately, theory and practice aren't always the same. No one has yet been able to build a diesel that performs in the way theory predicts. In reality, diesel cars emit about 10 per cent less carbon than petrol equivalents but are much 'dirtier', producing particulates and other pollutants than are implicated in asthma, lung disease and heart attacks. For greenness, stick with a fuel-efficient petrol model.

BIOFUEL – HANDLE WITH CARE

It sounded like the environmentalists' dream. A renewable fuel that could be grown and would thus be both inexhaustible and carbon neutral. The reality is very different. Using land to grow biofuels is driving up food prices and causing the clearance of millions of hectares of forest in Malaysia, Sumatra, Borneo and elsewhere. The demand for biofuels has already pushed the orang-utan close to extinction and threatens many other species including the Sumatran rhino, tigers, tapirs and various monkeys. There may be a case for some biofuels but they're just not the solution to the problem.

AND WHAT ABOUT GAS-POWERED CARS?

Some major manufacturers are now producing bi-fuel cars, that's to say, cars that can run either on petrol or liquefied petroleum gas (LPG). And most existing petrol cars can be converted. You can change fuels at the touch of a switch, which you may need to do fairly often since there are only about 1,200 LPG stations in the country at the moment.

The advantages are that LPG will save about 40 per cent on your fuel costs and 20 per cent on your carbon dioxide emissions, compared with a petrol car. Compared with a diesel, the carbon dioxide saving is only around 1.8 per cent, but the LPG car is significantly cleaner in terms of particulates.

The conversion, including an extra fuel tank, costs about £1,500 to £2,000 but you could recover that fairly quickly, especially if you use your car in London every day. As with all cleaner cars, you'll be exempt from the congestion charge which could add up to almost £2,000 a year.

The carbike – the real hybrid

We thought that might get your attention. Actually, there's no such thing as a car that's also a bike. But there are cars with bike racks. Basically, once you've got to your destination, you park your car for the rest of the holiday and use bikes for getting to the beach, doing your shopping and sightseeing in the area. Bearing in mind that cars are extremely inefficient for short hops, you'll save more fuel than you imagine, get fit and have a lot of fun.

THE FUTURE OF MOTORING HOLIDAYS

Despite some impressive investment, cars powered by hydrogen fuel cells are still a long way off. There are daunting technical problems to be overcome. Electric cars, on the other hand, are a practical proposition right now. The problem is only one of range. But all we need is a little imagination. Here's how it could work. You drive a couple of hundred miles and notice your batteries are low. You pull into the nearest 'electricity station' where a machine automatically lifts out your spent batteries and replaces them with a fully charged set. Off you go again. Meanwhile, the discharged batteries are recharged. And where would that power come from? To keep carbon dioxide emissions close to zero it would have to come from nuclear power stations, windfarms (offshore, to preserve the environment) or solar panels.

FREEWHEELERS

So a fuel-efficient car is relatively green if it's full. But supposing it's not? The solution is to contact Freewheelers (www.freewheelers.com) a car-sharing scheme set up to reduce pollution. It's a simple idea that's already well-established in Germany (they call it *Mitzfahrzentralen*), France (*Autostop*), Belgium (*Taxistop*) and Switzerland (*Sharcom*). All you have to do is register on the site, enter the details of your proposed journey (as far in advance as possible) and wait for someone to contact you. If you like the sound of them you offer them a lift on a cost-sharing basis. It's a win-win situation. You save money. Your passenger or passengers save money. And the environment gets saved. What's more, it's a great way of meeting people and passing the journey in conversation.

Freewheelers can also work the other way round, of course. If you'd like to go somewhere but don't have a car – or would simply prefer to go with someone else – register on the site and wait for someone to contact you.

Freewheelers is free – but does ask for donations.

Trains

Back in March 2007, Greenpeace activists at Gatwick airport were handing out free train tickets to Newquay. They were trying to convince people not to take BA's new 280-mile service to the UK's premier surfing resort. Why? Because the round trip by plane would generate 135kg of carbon dioxide as compared with just 36kg for the train. In other words, the train represented a saving of 99kg of carbon dioxide per person.

In fact, only one passenger took the free train ticket – probably because the flight time was only one hour compared with five for the alternative. But when you add in the time actually *getting to* the plane there probably isn't much difference for most people.

Environmentalists enthuse about trains. But are they really as green as they're claimed to be? The truth is some are but some aren't.

In the first place, railway lines require huge amounts of land. A ball park figure for a twin-track railway is three hectares per kilometre, including associated infrastructure. Take the train to the south of France and you're participating in the destruction of about 3,500 hectares of countryside. By comparison, Heathrow Airport, from which you can fly to literally hundreds of destinations, leaving the countryside below untouched, encompasses 1,200 hectares. So you could certainly argue that flying is less damaging to the landscape. On the basis of land use per million passenger kilometres, studies suggest that air and rail are about the same but air travel doesn't require building anything between airports. If you were to build a motorway with the same capacity as a railway or an airport you'd need two to three times more space. In terms of landscape, then, flying is best, followed by train, with road the worst culprit.

Then there's the cost. The French built their *Atlantique Train à Grande Vitesse* (TGV) line at around £15 million per kilometre. Their *Mediterranée* cost more than double that. And a possible bullet train line from Edinburgh to Glasgow has been costed at double again. So we're talking huge sums of money that could have been spent on other things.

What's more, as the speed of trains increases – to compete with planes

– so energy consumption rises significantly. In other words, trains are only more energy efficient than planes (or even large, three-quarters empty cars) when they're chug-chug-chugging along. Once trains get to a speed of about 350 kilometres an hour they use as much energy per seat as a passenger jet. Okay, you won't be rolling along at that kind of speed any time soon between London and Cornwall. But the TGV routinely achieves average speeds of 260 kilometres an hour in normal service and it has achieved a maximum speed of over 500 kilometres an hour in tests.

To be really, really green a train has to run on clean, renewable electricity, like the French TGV. If you take the TGV you can hold your head up because it's effectively nuclear-powered (always assuming, of course, you agree that nuclear power is 'clean'). But if the power comes from an unscrubbed coal-fired power station a train is not quite so environmental when compared with a car. Compared with a plane, however, it's always better in terms of pollution.

HOW PRACTICAL IS LONG-DISTANCE TRAIN TRAVEL?

The idea of travelling a long way by train can seem pretty daunting. For a start, there's the time factor. By plane you could be in, say, Marrakesh a few hours after leaving home. By train it's going to be more like two days. Then there are all those changes to negotiate. How are you even going to get all the information?

The time problem is solved fairly simply. Just think of the journey as an enjoyable part of the holiday. Which it can be. It's simply a question of mental attitude. You can have a nice meal in the restaurant. You can sleep overnight in your own private compartment and wake yourself up in the morning in your private shower. Not all trains are that way, of course, but plenty are.

The second problem has been best solved by one Mark Smith who, as a hobby, has set up a website called *The Man In Seat Sixty-One*. Quite frankly, it beats the operators' official websites hands down. You want to know how to get to Morocco by train? He'll give you every detail. The Far East? No problem. He's not even daunted by Australia. The answer to just about every question you ever thought of is here: **www.seat61.com**

CHEAP TRAIN TICKETS

Before you buy a train ticket you need to do a little research. Take a look at a website called MoneySavingExpert [www.moneysavingexpert.com] as well as www.thetrainline.com (domestic journeys) or www.internationaltrainline.com (international journeys).

Next, consider a Railcard. Railcards come in five categories:

- **Young Persons Railcard**. Ages 16-25 (26 if you're in full-time education). Saving: one-third. Cost £20.
- **Family Railcard**. For up to four adults and four children (5-15). You don't have to be related. Saving: one-third on adult fares, 60 per cent on children's fares. Cost £20.
- **Senior Railcard**. For those 60 or over. Saving: one-third. Cost £20.
- **Network Railcard**. For those travelling for leisure in South-East England. Saving: one-third on adult fares, 60 per cent on children's fares. Cost £20.
- **Disabled Person's Railcard**. If you're disabled you and one adult can save one-third of the fare. Cost £18 (£48 for three years).

It comes down to this. If you're making just one trip and the ticket costs more than £60 it's worth getting a Railcard. If you're travelling frequently by train it's essential. Note there are some restrictions but if you don't have to get somewhere early you'll probably be able to claim your discount.

You can even sometimes get first-class tickets quite cheap. They certainly don't have the sort of premium you'd have to pay on a plane.

Different companies have different policies but, in general, the longer you book in advance the cheaper the first-class ticket is likely to be. Most companies operate a three-tier structure. When the cheapest is sold out so the price jumps to the next level.

In reality, the cheap standard-class tickets tend to sell out quickest. Which means that when standard tickets have jumped up to the middle price, first-class tickets may still be at the cheapest level. And the price difference may then be quite small. In some cases, the extra cost of first-class can be as little as 15 per cent. At weekends, when business travellers are at home, upgrading to first-class can be a real bargain. It can add as little as £10 to the cost of a ticket from London to Birmingham and just £20 to the cost of getting to Glasgow.

What do you get for your money? That varies. More comfort, certainly, and some operators, such as Virgin Trains, offer complimentary food and drink (drinks and biscuits only at weekends).

THE FUTURE OF TRAINS

No matter what the environmentalists say, people don't like communal travel. They'll put up with it, of course, when they have to. But the truth is that people like cars. They like travelling in their own private space, together with family and friends, listening to their favourite radio programmes or music or whatever.

So where does that leave mass transport systems? The ideal, instead of travelling on a coach with 40 other people or a train with a few hundred (and that's just one carriage) would be a sort of pod you could step into near your house and get out of near your work. And that's exactly the idea of PRT or personal rapid transit.

It's not so far fetched at all. If you've ever been skiing and ridden up the mountain in one of those little gondolas then you've tasted a sort of personal rapid transit. And if you've been to Rotterdam recently or Schipol airport you might have ridden a PRT-like vehicle on the road. Quite soon, if you're using Heathrow's newest terminal, you'll be able to sample the real deal – a pod to get you to your car park.

There's no reason the idea couldn't be applied to everyday commuting. It would be far more convenient. Because pods are lighter than trains the rail is lighter, too – and smaller. Which means rails could be built in many more places. You'd get on at a station near you, programme your destination and automatically be whisked up onto the main line to join all the other pods. When you got near your destination your pod would divert itself into the station, leaving the main line clear. So there wouldn't be any of that tedious stopping at stations you didn't want to get off at. And the whole thing would be powered by electricity – hopefully from renewable sources.

It only needs vision to put PRT into practice. The scientists already have it. Now it's over to the politicians.

Boats

In 1866, 16 clippers gathered in the Chinese port of Foo-Chow-Foo, each captain determined to be the first back to Britain. Imagine something like today's Beaujolais Nouveau Race only with tea. The winners of the 15,000 mile Great Tea Race, as it was known, were Ariel and Taeping, arriving in London in 90 days. Three years later an even faster sailing ship was launched. She was capable of 17 knots and in one 24-hour period covered 363 nautical miles. She was, of course, the Cutty Sark.

The Cutty Sark wasn't just ahead of her time, she was ahead of *our* time, travelling the world on renewable energy without producing any carbon dioxide. One-and-a-half centuries later, with the benefits of modern materials and skills, why can't we match that? The reality is that shipping is now responsible for at least two per cent of greenhouse gas emissions and, according to some sources, the contribution to carbon dioxide emissions could be as high as five per cent. In other words, ships are, overall, just as bad as planes. And on a per passenger basis they're worse. So it's no good taking a cruise instead of a beach holiday, nor crossing the Atlantic aboard a liner.

It's been calculated, for example, that taking a boat from London to Naples would generate seven times more carbon than flying. Of course, the boat has much further to travel than the plane, but even on a direct route a ship is an energy glutton. A return trip to New York aboard the QEII generates around nine tonnes of carbon dioxide per passenger – almost eight times more than flying. Another way of putting it is that taking a holiday on a cruise ship is equivalent to flying 500 miles every single day of your holiday. Unfortunately, there are now something like 300 large cruise ships, carrying several thousand passengers apiece – a total of over 16 million a year.

But there is another way. The Cutty Sark way. The pity is that sailing ships have already been invented. Because if they were invented today they'd be hailed as one of the greatest technological advances of our era and everyone would be rushing to build them. And, with modern knowledge, they'd not only harness the wind but use wave power, too, to generate electricity.

A few companies have had that foresight. In 1986 the first commercial sailing vessel in 60 years was slipped at Le Havre and named Wind Star. Wind Song followed in 1987 and Wind Spirit in 1988. The idea was to build modern cruise liners that nevertheless carried real sails. At the touch of a button they unfurl in two minutes – no climbing the rigging. Of course they don't replace the engines but they make a significant contribution.

With modern technology improving on even the Cutty Sark's times, a wind-powered ferry could cross short stretches of water like the Channel in not much more time than a conventional ferry. Would an extra half-hour matter so much? What's more, on a cruise, speed is irrelevant. It's the experience of cruising that counts. And that would be all the greater on a sailing ship. If you want proof, turn to and read all about cruising the Med aboard Sea Cloud, and even sailing round the world.

For the time being there are no wind-powered ferries across the Channel or anywhere else. But if you can't use a ferry or a liner and you can't fly [*see below*] how are you going to get somewhere like the USA?

We have a couple of plans. The first, we admit straight away, is not for everybody. But then, you're not everybody. Here it is. You can sail across the Atlantic (and lots of other places) as working crew aboard a yacht. If you actually do know how to sail a yacht that's a big help. But even if you don't you're not ruled out. The thing is, there are large numbers of people who need to move a yacht from A to B (usually from the Med to the Caribbean or back) and who are just too busy making money to do it themselves. Or who have the time but just don't have any friends who have the time.

In the first scenario the owner appoints a captain and the captain finds a crew. In scenario B the owner is the captain. Either way, if you're capable of following some simple instructions and appear to be a congenial sort of companion for the hours you're all going to be crammed together then you're in with a chance.

In fact, although yachts aren't big it's easier to keep out of the way of other crew members than you might imagine. There'll be a watch system and when you're off watch you'll be sleeping. When you're on watch everybody else will be sleeping.

So how can you get into this? There are agencies that specialise in putting owners, skippers and crew together. In some cases you may have to pay a fee to register. Here are three of them:

- **Crewseekers**: www.crewseekers.co.uk Tel. 01489 578 319
- **Professional Yacht Deliveries**: www.pydww.com Tel. 01539 552 130
- **Reliance Yacht Management**: www.reliance-yacht.com Tel. 01252 378 239

The second idea still requires a degree of fitness not to say fortitude. But it's less demanding than the first option. It's to pay for a passage aboard a tall ship. Yes, they still exist as commercial vessels. Take a look at www.classic-sailing.co.uk and you'll see for yourself (Tel. 01872 580 022). Of course, these ships aren't plying back and forth on a regular basis but if you're in no particular hurry you should be able to find a crossing to suit. We'll have more to say about tall ships in *Part 2*.

You've probably got two big questions. How long does it take? And is it dangerous? As to the first, well, we're talking 3,500 miles or so, depending on the exact route. Think in terms of a month. As to the second, *Classic Sailing* recommends you to be fit and under 65. Where small yachts are concerned you also need to check out both the skipper and the boat very carefully.

CAN'T THEY MAKE BOATS MORE ENVIRONMENTALLY FRIENDLY?

Wind power aside, there are a couple of other possibilities. Nuclear power and hydrogen.

Nuclear power works for several of the world's navies, so why not commercial and cruise ships? Back in the 1960s there *was* a nuclear-powered cargo ship. She was the N.S. Savannah but, at a time when oil was cheap, she wasn't economical. Today it would be a different story.

As for hydrogen, the 130-ton Icelandic whale-watching ship Elding should, if all has gone to plan, now be taking tourists to see blue whales without creating any pollution. But we're not going to be seeing many more hydrogen boats any time soon. Hydrogen takes up a lot of space and most ships would only be able to carry enough for about a week. That's fine for the Elding but it's not fine for most of the world's shipping.

Planes

Flying is bad for the environment. There's no getting away from it. And it's not just a question of energy consumption and pollution emissions per passenger kilometre. They're serious enough but the real problem is that aircraft encourage us all to travel *more* kilometres than we ever would have dreamed of. So it's a case of more pollution per kilometre *and* more kilometres.

Would you, for example, ever have contemplated a weekend shopping trip in Rome if you could only get there by rail? Would you ever have taken a holiday in the Caribbean if it took a month to get there by sailing boat? Would you be thinking of a holiday Down Under if planes hadn't been invented? In fact, a holiday in Australia by air will produce as much carbon dioxide as you'd normally create *in an entire year*. From *every* activity. Think about it. A whole year's carbon dioxide in one holiday.

How serious is this question of carbon dioxide? It's an important subject where travel is concerned and we're going to be looking at it in the next chapter.

But for the moment we'll just tell you what some of the most concerned experts are saying. They argue that we need to reduce our carbon emissions to *one-tenth* of what they are now. They say it's not enough to hold them at the current level. It's not enough to cut them a measly few per cent nor even a whacking 50 per cent. They want cuts of 90 per cent. Which means that even a return flight to Spain would take a whole year's 'allowance.'

For them, there's simply no getting away from it. Green travel means *not* flying at all unless there's some extremely compelling reason to do so. In of this book we'll be telling you all about marvellous holidays that *don't* involve flying. And we'll be telling you about a few for which, in our opinion, flying can be justified. But only a few.

SO WHAT'S THE PROBLEM?

Getting a big aircraft into the sky and keeping it there requires a lot of energy. To give you an idea, a Boeing 737 burns something like 2,000 litres of kerosene on a single 250-mile flight. That's probably about as much fuel as you put into your car in an entire year. On a 1,000-mile flight it burns around 6,000 litres. That's a lot of fuel and, if your maths is good, you'll notice that short hops burn proportionately more than longer ones. That's because the whole taxiing and take-off thing uses a disproportionate amount of energy. Which is why environmentalists are so against jet travel for short distances. On a Boeing 737, for example, you need to fly over 1,250 miles before you reach peak efficiency.

Basically, flying uses something like five times as much energy per passenger mile or kilometre as the average car with its average number of passengers, or a coach, or a train. That means five times more pollution. The figure is very rough since there are so many variables. But that gives you a broad idea.

Aside from carbon dioxide, those pollutants include nitrogen oxides, sulphate aerosols, soot and water. The sulphate aerosols and the soot roughly cancel one another out – the former tends to cool the planet while the latter heats it. As for the nitrogen oxides, they form ozone in the presence of light. Ozone has the useful property of destroying methane, a greenhouse gas, but is itself a greenhouse gas, and the net effect is negative. Finally, water vapour is, believe it or not, as harmful as carbon dioxide *when it's released at high altitude*. If there's a contrail – a cloud-like trail behind the aircraft – then it's harmful.

But won't improving fuel efficiency solve the problem?

To be fair to the airlines, fuel efficiency has been improving at something like two per cent a year. That may not sound a lot but over a decade it adds up. Boeing's new 787 Dreamliner hadn't entered service at the time of writing but it was being claimed it would prove to be as fuel-efficient as a car. (An American car, presumably.) And the generation after *that* could see carbon dioxide emissions cut by half. So flying is narrowing the once substantial gap with land transport.

Come on, aviation accounts for only two to three per cent!

Yes, but air traffic is growing so fast it's become a question of running to stay still. And, unfortunately, the aircraft designers just can't run fast enough. Like someone on the wrong escalator, the airline industry is going backwards when it comes to total emissions. True, aviation does only account for around two to three per cent of global output of carbon dioxide from man-made sources. But according to the report on climate change by Sir Nicholas Stern carbon dioxide emissions from aircraft could triple by 2050. That's serious.

Nevertheless, it's worth reflecting on where all the other emissions come from. In fact, road transport is responsible for almost a fifth and other forms of transport for around four per cent. Industry generates almost another fifth. In our homes we're responsible for almost 10 per cent. And the generation and use of energy outside the home accounts for about 40 per cent. So there are plenty of targets in addition to aviation.

THE FUTURE OF AIR TRAVEL

The next generation of aircraft – that's to say, being built in 2015 – could have twin engines with open rotors set just above the rear fuselage between a sort of double tail. It's a design that could improve fuel efficiency enormously, halving carbon dioxide emissions and also producing 75 per cent fewer nitrogen oxides. EasyJet is already pushing Airbus and Boeing to start building. But there's a problem with those kinds of engines – they're likely to produce a good deal more noise. And, in any case, a 50 per cent reduction in carbon dioxide emissions, however impressive, just isn't enough.

Until someone comes up with a way of flying that doesn't use such huge amounts of energy and doesn't produce such large amounts of carbon dioxide we're all going to have to rethink our holiday travel. And the truth is that no one has yet come anywhere near.

Richard Branson is throwing a lot of money at the problem. All credit to him. Virgin Atlantic and Boeing together were planning to demonstrate a biomass fuel on a 747 widebody around the time this book was published. But even if it succeeds, there simply isn't enough space on the planet to grow enough for our energy needs, let alone feed us as well.

To give you an idea, one source of palm oil is Riau province, Indonesia, where forests are being cleared for palm oil plantations. But apart from the impact on wildlife, the logging, burning and draining could release as much as 14.6bn tonnes of carbon, according to a Greenpeace estimate. That's equivalent to one year's greenhouse gas emissions. It's a disaster, which is why Greenpeace has been blockading biofuel tankers.

Maybe governments could agree that only airlines can use biofuels. Everyone else would have to find a different solution. But that's not going to happen.

Another line of research is to use liquid hydrogen as a fuel. It could be burned in jet engines similar to those used today and the only waste product is water. Perfect? Unfortunately not. As we've already seen, water vapour isn't a benign substance at the altitude at which planes normally fly. The net result could be to make global warming worse, not better. What's more, the creation of hydrogen, the compressing of it, and the maintaining of it at the necessary minus 259 degrees, would all use such formidable amounts of energy as to make it an almost pointless exercise. Unless somebody can come up with a way through these formidable technical problems, hydrogen just isn't going to be the solution.

Or is it? Put it into a balloon instead of an aeroplane and you have a quite different set of figures.

Don't laugh. Yes, OK, at the moment airships can only fly at a maximum of 130 kilometres an hour (about 80mph) so a journey from London to New York would take almost two days. And if the wind was in the wrong direction it would be longer. But, in fact, that's not as risible as it might seem. The airship would be part of the whole holiday experience. And a stunningly beautiful one at that. Out of a two-week holiday, four days watching the sunsets over the Atlantic would be a pretty romantic way of doing things. And, of course, there could be plenty of diversions on board. We'd happily travel that way.

An organisation known as the Tyndall Centre for Climate Change Research has estimated that the total climate impact of airships would be only 10 to 20 per cent that of jets.

So get down to your travel agent today and ask them: "When's the first airship leaving for New York?"

IS FLYING EVER JUSTIFIED?

Remember, we can't be concerned here with your personal timetable. We're concerned with something far more important – the environment we all depend on. So the fact that you're in a hurry isn't, unfortunately, a valid reason for flying.

But here, in our opinion, are some of the reasons flying might be justified:

✓ If you're making a major financial contribution to environmental projects in the destination by, for example, paying significant entrance fees to game parks.

✓ If you're playing a significant role in a genuinely important environment project [see some of our suggestions in Chapter 11].

✓ If you're visiting a Third World country where tourism is significantly improving the living standards of local people.

✓ If you've 'saved up' carbon offsets by modifying your behaviour at home.

We'll be discussing carbon offset schemes later. Some environmentalists are completely opposed to them, calling them 'red herrings'. They argue that such schemes encourage people to carry on flying. We take a different view. We believe you should only fly if you have a justifiable reason, such as those cited above, and then you should pay for carbon offsets *in addition*. Yes, in reality, carbon offset schemes fall far short of mopping up the amount of pollution they need to in the time available.

They're not the solution, it's true, but, nevertheless, they're part of the solution.

Nuclear-powered travel?

Most environmentalists have been against nuclear power. The environmental impact of Chernobyl was awful enough and another major accident could be catastrophic. Nor, even after all these years, has anyone yet built a proper long-term storage facility for radioactive waste, let alone found a way of making it safe. Nevertheless, increasing numbers of environmentally minded people are coming round to the view that nuclear power is the only solution to energy problems. Wind, wave and solar power can help but they just don't seem capable of meeting all our needs.

How could nuclear power solve travel problems? The most obvious way is in producing electricity for electric cars and trains. Cars would have battery packs that could be swapped at garages. The empty batteries would then be recharged with power from the nuclear power station. And some trains – for example, the French TGV – are already running largely on nuclear-generated electricity. Nuclear-powered cruise ships could be built.

Some critics say that, potential dangers aside, there just isn't enough uranium in the world to provide our energy needs. That's wrong. Uranium is widespread. It's even in seawater. It's simply a question of price. At the low prices seen in the 1990s uranium was too costly to extract from any but a few locations. But at the higher prices as this book went to press the economically viable deposits are sufficient for a few hundred years.

Getting across the Channel

Great Britain is an island. You'd probably noticed that already. Which complicates the whole matter of going abroad. So what's the greenest way of getting across the water? By air, by ferry or by the Eurotunnel? Plainly, it's not air. Nor indeed is it the ferry. You can argue that the construction of the Channel Tunnel released vast amounts of pollution.

But since it already exists, the relatively small amount of extra pollution created by the trains makes this the green choice. So whether you're travelling onwards by bike, coach, car or train opt for Eurotunnel. Unless, of course, you're a very good swimmer …

Chapter 2

Travel and climate change – the truth

FACTS YOU NEED TO KNOW

- Global warming is a fact not a theory – if it wasn't for global warming the Earth would be uninhabitable.
- For centuries the degree of global warming was relatively stable.
- In the last few decades global warming has increased – and at an accelerating rate.
- How far you travel on holiday could affect global warming.
- Where you go on holiday will be impacted by global warming.

As we saw in the last chapter, the part of your holiday that's likely to consume the most energy is getting to and from the destination. Why does that matter? Because, in the first place, when you burn fossil fuels you're depleting a resource that's in increasingly short supply. And, secondly, burning fossil fuels also creates pollution.

There's been a huge amount of controversy about the role of pollution in climate change. Here we're not going to give you opinions. We'll just give you the facts and you can decide for yourself.

Is there global warming?

The first thing to sort out is whether or not there actually is any global warming so far. Well, as the table below shows, the five warmest years in more than a century have all occurred during the past decade or so.

The five warmest years worldwide since the 1890s

1 2005
2 1998
3 2002
4 2003
5 2004

Source: NASA

Of course, that doesn't tell us what caused the warming. And it certainly doesn't prove that the warming will continue. But, unless you dispute the figures, it tells us that *there has been a period of warming.*

What kind of temperature increases are we talking about? Well, the Intergovernmental Panel on Climate Change (IPCC), created in 1988 by the World Meteorological Organization and the United Nations Environmental Programme, found that global average surface temperatures increased by 0.6°C (plus or minus 0.2°C) over the 20th century. That may not sound a lot but, as we'll see in a moment, it's highly significant. But the most ominous thing is that over the past 30 years the pace of warming has accelerated.

Aren't these temperature measurements inflated by 'urban heat islands'?

Scientists can make mistakes just like anybody else but that one is just too obvious. The higher temperatures in cities – your so-called urban heat islands – are discounted when the average surface temperatures are calculated.

It's important to recognise that these are average figures. Some parts of the globe have warmed more and some have warmed less. Over the past 50 years the largest annual and seasonal increases have occurred in high latitudes, especially Alaska, Siberia and the Antarctic Peninsular.

The second warmest year of the past century – 1998 – occurred when there was a strong El Niño effect. So that might explain that one. But the warmest year so far happened without the help of El Niño.

What's El Niño?

El Niño is a regular though infrequent climate quirk which increases the surface temperature of the tropical Pacific. Amazingly, this in turn can lead to wind changes throughout the whole world causing hurricanes, flooding and drought.

The opposite is La Niña, a less common cooling effect which, while it reduces extreme events in the Gulf of Mexico and the Caribbean, can nevertheless lead to increases in tropical storms and hurricanes in the Eastern Pacific.

Is there any link with global warming? The jury is out on that one. But it's for sure that the droughts caused by El Niño lead to brush and forest fires (especially in Australia). In turn, those fires both produce carbon dioxide and reduce the planet's ability to absorb it.

Will the global warming continue?

The only way to predict the future is to use climate models. Different scientists use slightly different models and come up with slightly different results. The range of mainstream predictions at the moment is that temperatures will increase by between 1.4°C and 5.8°C over the next 100 years – that's roughly two to ten times more than in the past 100 years.

How good are these climate models?

At the moment the perfect model doesn't exist, but the scientists realise that, of course. That's why they always give a range of possible error with their predictions.

A good model would be one that if you fed into it the climate data for, say, 1990 and then asked it to forecast 10 years ahead, it would come up with exactly the figures that we know with hindsight were the correct ones for 2000. Which is exactly what Doug Smith and his colleagues did at the Hadley Centre in Exeter. So the Hadley Centre has a pretty good forecasting programme.

And what it tells us is this. The years 2008 and 2009 will be relatively cool (like 2007) but between 2010 and 2014 three years will set new records.

Won't global dimming counteract global warming?

Gerald Stanhill designed irrigation systems in Israel. Not just any old irrigation systems. His systems delivered the perfect amount of water. Enough for the plants to thrive but not one wasteful drop more. And for that he needed to know exactly how much sunlight fell on the plants. In the first few years he noted nothing odd but as the years turned to decades he began to notice something strange. The amount of sunlight falling on Israel was going down. Together with a colleague, Shaptai Cohen, he published a scientific paper describing the discovery of what they called 'global dimming'.

Global dimming is one of the great unknowns in the climate debate. The problem is we just don't have enough radiometers – light measuring devices – set up around the world to know exactly what's going on. But where there are reliable records they all point to a 10 per cent reduction in the amount of sunlight reaching the Earth. And that's in just 30 years!

What's going on? One plausible theory is that pollution is putting aerosols into the atmosphere, which reflect sunlight back into space. That fits with some of the data but not all of it. For example, it would explain why temperatures didn't rise as much as they should have done during the 1950s, 60s and 70s. But China is a country that *does* have solar radiation figures and they show a 'global brightening' since the mid-1980s, despite China's vastly increased air pollution.

So the picture is far from clear. But whatever is going on with global dimming, global warming is carrying on regardless.

OK, but where's the evidence that global warming is due to the greenhouse effect?

When we first heard about global warming we asked the same question as everybody else. Is this for real? Then we discovered something that puts the whole subject into perspective. The greenhouse effect is definitely not a theory. It's not new. And contrary to what you may have heard, it's not disputed by any scientist at all. The fact is the greenhouse effect has existed for eons and without it the average temperature of the planet would now be minus 18°C. In other words, about 33°C lower than it is now. We wouldn't be writing this and you wouldn't be reading it, if it wasn't for the greenhouse effect. None of us would exist. Thank goodness for it.

The greenhouse effect works like this. The main constituents of the atmosphere – nitrogen and oxygen – are transparent to heat radiation. If they were the only gases in the atmosphere, the heat generated on Earth by the sun would be radiated back into space. But, fortunately, there are additional gases present in small quantities – water vapour, carbon dioxide, methane and others – which can trap some of the heat. And it's lucky for us that they do. You've probably noticed the effect for yourself. When the sky is clear the night-time temperature falls more than when it's cloudy. That's because the clouds – water vapour – are exerting their greenhouse effect.

All this has happened naturally – that's to say, without any man-made intervention – which is why scientists call it the *natural greenhouse effect*. As we stress, it's *not* a theory. The ability of carbon dioxide, methane and various other gases to absorb long-wavelength radiant heat energy emitted by the Earth has been demonstrated in practical experiments.

Essentially, greenhouse gases operate like insulation round the Earth. Think of it like this. When you turn on a heater in your home the room warms up. When you turn it off the room stays warm for a time because the walls stop the heat escaping. That's the same as the greenhouse effect on the planet. Now imagine taking the heater out into the garden on a cold night. Stand in front of it and you're warm. But the moment you switch it off you're instantly cold. Why? No insulation.

We can see the mechanism at work all around us in the universe. To give you an idea, the atmosphere of Venus, which is almost exactly the same size as the Earth, is composed *almost entirely of carbon dioxide*. So if the global warming theory is correct Venus should be an extremely hot place. And, you know what ... it is. It's 460°C, in fact. No life, as we know it, could exist there.

Aha! say the sceptics. You're overlooking the fact that Venus is much nearer the sun than the Earth is. At first sight, it seems like a reasonable point to make. But then take a look at Mercury. Mercury is very roughly half as far from the sun as Venus and so if distance from the sun were the only factor at work it should be hotter still. But the temperature of Mercury is very curious. On the sunny side, yes, it's hot but *not* as hot as Venus. It's 427°C. While on the night-time side the temperature is ... wait for it... *minus* 173°C. Why? You've guessed it. Mercury has no carbon dioxide. In fact, it has no atmosphere at all. So there's nothing to trap the sun's heat. Where it shines, the planet rapidly heats up. Where it doesn't shine, the planet rapidly cools down. Comparing Venus and Mercury gives a pretty clear picture of carbon dioxide at work.

Your own greenhouse gas experiment

If you've never particularly noticed the effect of night-time cloud cover then now is a good time to do a little experiment of your own. All you need are:

- An atmospheric thermometer with maximum and minimum readings
- A notebook

Set out your notebook like this:

	Day 1	Day 2	Day 3	Day 4	Day 5 …
Maximum daytime temperature					
Minimum night-time temperature					
Temperature difference					
Night-time cloud cover					

(*for example, none, light, thick etc*)

You'll find that when the skies are clear at night the temperature difference between night and day is generally greater than when there's thick cloud at night. You'll have proved for yourself that the greenhouse effect exists. In this case it's due to water vapour. But carbon dioxide, methane and certain other gases work in the same way.

So far so straightforward. The pre-industrial concentration of carbon dioxide in the Earth's atmosphere was 280 parts per million (ppm). Today the level stands at 380ppm. Those who dispute human influence on global warming have got to explain why carbon dioxide *does* warm the planet at concentrations up to 280ppm but exerts no further warming effect at higher concentrations. That's a difficult trick to pull off. It would be the same as trying to argue that one centimetre of insulation in your home keeps the temperature up but two centimetres doesn't make any difference. It's just not logical.

But how do we know carbon dioxide is responsible?

So far we've concentrated on carbon dioxide but, yes, there are other gases that trap heat. Here are the main ones, compared with carbon dioxide:

	Lifetime in the atmosphere	Pre-industrial concentration	Concentration in 1998
Carbon dioxide	5-200yrs	280ppm	365ppm
Methane	12yrs	700ppb	1,745ppb
Nitrous Oxide	114yrs	270ppb	314ppb
Chlorofluorocarbon-11	45yrs	0	268ppt
Hydrofluorcarbon-23	260yrs	0	14ppt
Perfluoromethane	50,000yrs	40ppt	80ppt

ppm = parts per million; ppb = parts per billion; ppt = parts per trillion

As you can see, carbon dioxide is the most abundant greenhouse gas but it's not, in fact, the most powerful. If its *global warming potential* is given a value of 1 then, over a 20-year period, methane is 62 times stronger and over the course of a century nitrous oxide is almost 300 times more powerful. However, because of its abundance, carbon dioxide is the single largest contributor to what scientists call *radiative forcing* – the process by which human activities intensify the natural greenhouse effect.

The most frightening figures are those in the first column. Even if all man-made emissions are stopped now – which would be impossible – the planet would still heat up because it takes years, and even centuries, for these gases to disappear from the air.

If that's right, why are so many people unconvinced?

In 2007, 315 financial institutions together controlling around $40 trillion in funds sent out a joint letter. It was addressed to the 2,400 largest firms in the world. Basically, the institutions wanted a straight answer to a very important question: What are you doing about global warming?

Now, the heads of these financial institutions are, no doubt, very nice, concerned people. But they don't normally waste their professional time on altruistic deeds or wild goose chases. Their job is to make money. So why did they put their names to the letter? Two reasons. Firstly, they believe man-made global warming is a reality. Secondly, they believe it could seriously impact their investments.

Of course, that itself doesn't make them right, and we'll be looking at some of the opposite scientific arguments in a minute. But that's the measure of how seriously the business community now takes global warming.

What about sunspots?

Sunspots show cyclical patterns of 11, 90 and 180 years. Scientists have known about them for a long time and during the so-called Little Ice Age (1650-1750) very little sunspot activity was observed. So there could be a connection between the Earth's climate and sunspots. But when you look at a graph of the Earth's surface temperature over the past 1,000 years the Little Ice Age appears quite trivial when compared with the much greater temperature change we're now experiencing. Sunspots just can't explain it.

What about changes in the Earth's orbit around the sun?

This is the so-called Milankovitch Theory. It proposes that cyclical variations in three of the Earth's orbital characteristics affect climate. The first is *eccentricity*, that's to say, the change from an elliptical orbit to a more circular one. But that takes place over a period of 100,000 years. The second is the wobbling of the Earth on its polar axis, a phenomenon more technically known as the *precession* of the equinox. But that cycle takes place over 26,000 years. It means that in 13,000 years from now, the northern hemisphere will experience greater seasonal variations – hotter summers but colder winters.

Then there are changes in the Earth's tilt or obliquity. But that cycle is 41,000 years.

So although all of these things undoubtedly affect the Earth's climate, none of them can explain the rapid increase in global temperatures over a timescale of a few decades.

But aren't there scientists who disagree?

George Monbiot has made an exhaustive study of this subject in his book *Heat* and we recommend that you read it. To summarise Mr Monbiot, many of the contrary opinions emanated from a body called the Science and Environmental Policy Project (SEPP) run by Dr S. Fred Singer. They were extensively quoted on the internet and then requoted by many journalists. But according to Mr Monbiot, SEPP was funded by ExxonMobil, the oil company, as were more than 100 other organisations, all following the line that 'the science is contradictory, the scientists are split, environmentalists are charlatans ...'

It seems to have been a very successful campaign of misinformation. Against Dr Singer and the small number of other contrary voices are ranged, for example, the many scientists whose research has been carefully examined by the Intergovernmental Panel on Climate Change (IPCC).

The IPCC was set up by the World Meteorological Organization (WMO) and the United Nations Environment Programme (UNEP) and it's important to stress that it doesn't carry out its own research but, rather, reviews and amalgamates the research by scientists all over the world. So it's not just one voice but the combined voices of thousands of scientists from every part of the globe.

It's worth pointing out that the IPCC won a Nobel Prize in 2007 for 'efforts to build up and disseminate greater knowledge about man-made climate change'. You can read the details of the IPCC findings at: **www.ipcc.ch/index.html**

So what's the impact likely to be?

The following changes won't happen instantly a certain temperature increase is reached. The melting of glaciers and icecaps, for example, is a process that will take many years. But the changes will then be inevitable.

ONE DEGREE OF GLOBAL WARMING

What it will mean for the planet

One degree would take us back to the way the Earth was 6,000 years ago. In the Western USA it would be worse than the dustbowl years of the 1930s, when thousands of American farmers and their families became refugees. By 2100 it would mean there would be no fresh water anywhere on one-third of the world's land surface. The melting of sea ice will accelerate, sea temperatures will rise, and hurricanes will increase in frequency and intensity.

What it will mean for your holiday

You might prefer to avoid destinations such as the south of Spain, Greece and Turkey in high summer – temperatures well into the 40s C (over 100°F) will be the norm. You might have to forgo low-lying destinations such as the Maldives – they won't exist any more. And you'll need a gondola to navigate not just the canals but also the streets of Venice.

TWO DEGREES OF GLOBAL WARMING

What it will mean for the planet

Coastal towns – Manhattan, London, Bombay, Shanghai, to name a few – will be flooded. You'll no longer need snowshoes for a visit to Greenland – it'll live up to its name and be green, not white. And India will be a drought zone.

What it will mean for your holiday

You certainly won't be taking Mediterranean summer holidays any more. Apart from the fact that it will be too hot, all the existing coastal infrastructure will become submerged. The Caribbean in the hurricane season will be a no-go area. And you can forget skiing.

THREE DEGREES OF GLOBAL WARMING

What it will mean for the planet

The last time the planet was this hot was about 3,000,000 years ago. Sea levels were then a staggering 25 metres higher than today, so today's coastal towns just won't exist and hundreds of millions of people will be displaced.

What it will mean for your holiday

You certainly won't need to leave the UK to get a suntan, to put it mildly. In fact, summer temperatures will consistently be in the mid-40s C (over 100 F). Nor will you need to travel to Greece to enjoy a collection of islands because that's what Britain will have become. Australia will no longer be on the itinerary, nor the game parks of Kenya or Tanzania. You'll probably be heading, instead, for the Arctic.

All right, but it's going to take a long time and by then we'll be able to go to another planet, won't we?

No to both questions. Sea levels won't rise 25 metres on the very first day that global average temperatures reach three degrees higher. But life could become very unpleasant as early as 2050 and we're talking in terms of sea levels rising a metre in the next century.

As to other planets, there are none in the solar system that are capable of supporting us, even if we had the technology. We've already ruled out Venus and Mercury. Mars looks more promising. Life on its equator wouldn't be too bad, varying between minus 20°C and plus 30°C. But as to the rest of it, well, the average surface temperature is minus 65°C. So not much room there. And all the other planets are colder still.

Earth alone in our solar system has a climate that can support us. And, in fact, when you think about it, only a small fraction of the Earth is actually comfortable. Seven-tenths of the planet is covered by water so we can't very easily live there. Of the remainder, vast areas are too cold, too icy, too hot, too desertified, too steep, too high or have too great a flood risk to make congenial places for humans to live. In other words, in the whole of the solar system there's just between one and four acres per person, to live on and to supply all our needs – houses, gardens, farmland, roads, railway lines, factories, shops and all the rest.

If that were to be reduced because of climate change it would be a catastrophe.

What sort of cuts in carbon dioxide emissions will we have to make to avoid all this?

As mentioned in the previous chapter, some experts are calling for cuts of 90 per cent. Which, with present technology, would mean things like an end to flying and almost completely banning private cars.

Is there another way?

This is where things get really controversial. We believe there is. For two reasons. Firstly, because the amount of carbon dioxide *already* in the atmosphere will mean global temperatures going up by between a half and one degree Centigrade. And as we've just seen, that will already be pretty serious. That scenario, remember, is with the carbon dioxide we already have. And more is being added to it every second. If we don't like the one degree scenario – let alone anything worse – we have to use our ingenuity, either to extract the carbon dioxide from the atmosphere or to offset its impact.

That brings us to the second reason. An alternative just *has* to be found. Because, no matter how sensible it might be from the environmental viewpoint (and, indeed, the personal survival viewpoint) to ban planes and severely restrict cars and so forth, it isn't going to happen. It just isn't. By the time we all accept the case for such severe changes in lifestyles it will be far too late.

We might be accused of wishful thinking but necessity has always been the mother of invention and it will be again. The race is on for new technologies.

One approach is to reflect some of the sun's heat back into space. How? Well, snow is pretty good at it. Which is why the loss of sea ice, glaciers and snow in winter is so serious. It's not just a question of no more skiing. All those chocolate-box snowscapes were actually helping maintain the planet at an agreeable temperature.

A technical way of achieving a similar result would be to fire tiny drops of seawater into the sky to create highly-reflective, low-lying marine cloud. Stephen Salter at the University of Edinburgh has designed an unmanned 'ship' to do just that, using nothing more than wind power. According to his calculations, 50 such ships, costing a few million pounds each, would take care of one year's carbon dioxide emissions. What they'd do to the weather isn't clear but building an experimental ship sounds like a good idea. And urgently.

Another idea is to spread tiny particles in the upper atmosphere. This already happens to some extent both naturally and unnaturally. The natural way is a volcano – like Mount Pinatubo – whose 1991 eruption caused measurable global cooling. The unnatural way is sulphate pollution from industry. Nobody wants pollution and in the last decade or so clean air legislation has reduced it. So it would be ironic if another form of pollution turned out to be part of the solution.

The most expensive and exotic idea so far is to launch tiny, reflective spacecraft – no more than a metre across – to the inner Lagrange point. That's a spot between the Earth and the sun where the forces should hold a spacecraft locked in position. Unfortunately, several million would be needed at a cost in the trillions of dollars.

Other more easily attainable ways of reflecting sunlight could include reflectors in the desert (possibly combined with solar cells) and, quite simply, painting the roof of every house and building white. Los Angeles is already seriously considering this, not just to help slow global warming but to reduce local warming – big cities get hotter than the countryside.

But perhaps the most logical approach is to remove the excess carbon dioxide from the atmosphere. After all, we put it in. So we should be able to find a way to take it out again. Algae are good at scrubbing carbon dioxide and could be encouraged to do more. One idea, already being tested, is to seed the oceans with iron, which stimulates the growth of algae. Another is to float 100 metre tubes vertically in the water, equipped with non-return valves. The movement of the tubes by the waves would draw cooler water up and improve conditions for the algae, which are now suffering in today's warmer oceans.

Yet another idea is to build structures looking rather like cooling towers. Upward-flowing air inside would pass through a cloud of sodium hydroxide, causing the carbon dioxide to fall to the bottom.

Have we reached peak oil?

Once we've passed peak oil so production levels will decline ... and decline ... and decline. Which means carbon dioxide emissions from that source will also inevitably go down.

In 2007 demand was around 86 million barrels *a day* and still going up, largely due to China and the other emerging economies.

But have we now reached peak oil? And will we all be forced to change our ways, whether we like it or not? Well, at the time of writing, oil prices have set new records and oil companies, big and small, are out there scouring the planet for new fields. But they're not finding them. New wells in existing fields, yes, but completely new fields are rare.

As a ball park figure, the output of an average oil field declines almost five per cent a year. US production, for example, is already past its peak. That was back in 1973.

Some experts say we're at the global peak now. Some say we're already past it. The most optimistic put it at 50 years.

The strange case of Robert Preston

It was one of the most astonishing cases in legal history. Robert Preston had been doing nothing more than standing in the high street of the small country town where he lived when a man approached, raised his arm and fired a shot. What happened next has been the subject of controversy ever since. Onlookers described seeing Mr Preston clutch his chest and fall to the ground. While an off-duty nurse tended him, three courageous bystanders heroically decided to have-a-go. But, the curious fact is, the gunman offered them no resistance and when the police arrived he even laughed.

Mr Preston died and the gunman was put on trial for murder. His defence was unusual. He didn't deny firing the gun. He simply argued that Mr Preston had died for a completely different reason. Mr Preston, so the gunman claimed, had died of a heart attack before the bullet hit him. Therefore, the shooting had nothing to do with it. It was merely a coincidence. He even called two eminent scientists in his defence and they confirmed the gunman's case. Meanwhile, although the forensic scientists called by the prosecution were initially adamant that Mr Preston died from the bullet, under fierce cross-examination, they all had to admit they couldn't prove beyond doubt that Mr Preston hadn't died of a heart attack before the bullet smashed into his chest. And, indeed, it emerged he had suffered heart attacks before.

When the jury came back the foreman, who was in fact the editor of a respected newspaper, pronounced: "Not guilty!" The courtroom was in an uproar. The judge then handed down his now famous ruling. Subversive forces, he said, wanted us all to believe guns were dangerous when there was insufficient evidence. Until such time as it could be conclusively proven that people who had been shot had not died of prior heart attacks, it would be permissible for anybody to shoot anybody else.

You think we're making that up, don't you? But we're not. All we've done is change the names. Instead of Robert Preston read "the atmosphere of the Earth". And instead of the bullet read "man-made carbon dioxide". Otherwise the scenario is the same.

Chapter 3

Putting right the damage

FACTS YOU NEED TO KNOW

- Carbon offsetting has been the Earth's natural way of making the planet habitable over millions of years.
- Nevertheless, some ecologists argue that "artificial" carbon offsetting is a dangerous idea.

On Monday 20th August 2007 a strangely-dressed group of people gathered outside the King's Cross, London offices of CarbonNeutral. They were, rather bizarrely, adorned as red herrings. Meanwhile, in Oxford, another group handed in a parcel at the offices of Climate Care. When it was opened it was found to contain, yes, herrings.

One of the activists explained to the press that red herrings symbolised the futility of the carbon offsetting the two organisations were offering. "We deeply believe you should be getting on a bike instead of getting in a car," she said, "and thinking about whether you need a foreign holiday."

A spokeswoman for CarbonNeutral responded that the company was on the "same side" as the protesters. "We do understand why they're protesting," she said, "we just think they picked the wrong people to protest against."

Without a doubt, trying to put right the damage to the Earth's atmosphere through carbon offsetting is controversial. The most idealistic environmentalists say that generating carbon dioxide unnecessarily is simply wrong. And you can't therefore offset it by doing some other good deed. As far as they're concerned, that would be like offsetting a murder by giving money to charity. So flying is out. In fact, in their eyes, carbon offsetting is actually *increasing* the production of carbon dioxide because it encourages people to believe it's all right to carry on as before. And anyway, they add, carbon offsetting schemes don't actually work. At best, they're a red herring, at worst they're positively dangerous.

All good, valid points. Except for one thing. They make the assumption that most people can be persuaded to give up flying. And all unnecessary journeys of whatever motorised sort. Well, the people who dress up as red herrings are obviously willing. And you might be willing, too. But, in that case, we believe you're very much in the minority.

We see things differently. We believe the world is going to continue to produce dangerous amounts of carbon dioxide. In fact, in the current situation *any* carbon dioxide is dangerous. And it doesn't just come from transport, of course. No matter how much carbon dioxide emissions are reduced through energy-saving schemes, lifestyle changes and all the rest, there's *still* going to be a need to do something more. And that something can include carbon offsetting.

It shouldn't be controversial. After all, it was carbon offsetting that made the planet fit for human beings in the first place. Algae in the oceans are a natural form of carbon offsetting. Trees are a natural form of carbon offsetting. Just about everything that lives and dies and gets buried is a form of carbon offsetting. So what's wrong with a little 'artificial' carbon offsetting?

What is carbon offsetting?

To recap, transport consumes huge amounts of energy and as a consequence produces huge amounts of pollution. One of the pollutants is carbon dioxide which, as we've already seen, is a greenhouse gas. And just to emphasis a point already made, that's fact. It's not a theory [see *Chapter 2*].

The idea of carbon offsetting is logical and simple. If your holiday produces, say, one tonne of carbon dioxide then you have a responsibility to ensure it's somehow reabsorbed. As a result, you can enjoy your holiday with a clear conscience, knowing you're not adding anything to the greenhouse effect.

So how can you achieve this offsetting? Broadly, there are two routes. Either you pay money to an organisation to do the offsetting on your behalf. Or you do it yourself. In our opinion, both methods are valid and we think you should do both together. Call it belt and braces.

Running your own carbon offset scheme

When you contribute to a carbon offset scheme you get a nice feeling but you don't know exactly how much good you're really doing. On the other hand, if you do it yourself you have a much higher degree of control. In any case, we should all be roughly aware of our 'carbon footprints' and constantly looking for ways to reduce them.

Let's say, for example, you're planning a holiday by air in Tunisia. As a result, you're going to have to offset around two tonnes of carbon dioxide. That's quite probably about one-fifth of your total *annual* production. Which just emphasises how energy-intensive holidays can be. So how could you find those kinds of savings in your daily life to offset your holiday?

There are whole books on the subject of green living and we suggest you read one. But here are some ideas to get you started.

SWITCH TO A MORE FUEL EFFICIENT CAR

Your motoring probably accounts for about one fifth of the carbon dioxide you create. So this is an important area to focus on. Let's say you currently drive a Land Rover Discovery or a Mercedes M class. So here's the deal. You can't take your holiday to Tunisia unless you swap it for something that's vastly more fuel-efficient. If you need a large saloon then you might go for the hybrid Toyota Prius, for example. If you do you'll be saving around 200 grams of carbon dioxide per kilometre.

Assuming you drive the average distance, that would amount to four tonnes a year. If there are normally two of you in the car, that's two tonnes apiece. So that's your holiday to Tunisia 'paid' for.

If you're actually going on holiday in your car you can see the saving more directly. Let's say you're driving down to one of the Spanish costas as a family of four. Call it 2,500 miles (4,000km) for the return trip. If you're in a large family saloon doing 20 mpg averaged out over the journey, that amounts to 1.2 tonnes of carbon dioxide. Which is 300kg each. But now suppose you do the journey in a Toyota Prius. You'd only produce half the carbon dioxide and possibly less. That would be a saving of 600kg. So far so good.

But you've still got to offset the 600kg you're still producing, of course. Which means each of you has to find savings of 150kg. If three times a week you each walk or cycle journeys of under two miles (which, statistically, are a quarter of all journeys) that would do the trick.

SAVE ENERGY AT HOME

Heating is the biggest single user of energy in anybody's home, accounting for around 70 per cent of the total. So that's the place to start.

One of the most dramatic things is to convert single-person households into two-person households, thus more or less halving energy use at a stroke. So, if you're one of those people who, up to now, has been wary of making a commitment then, go on, take the plunge. Tell him or her: "I really want to be with you ... and we'll save energy together."

Apart from that, you could:

- Turn your thermostat down one degree C – over a year that would be equivalent to one return flight to Belgium.
- Insulate your loft, walls (in the cavity if you have them, with external insulation if not), floors and water tanks – over a year that would be equivalent to one return flight to Germany.
- Switch to a renewable electricity supply. To find out what's available

in your area take a look at www.greenelectricity.org. Over a year that could be the equivalent to one return flight to Spain.

- Heat your entire house using a clean-burning stove and wood from a local sustainable source. Your carbon emissions will effectively be zero because wood is part of a continuous cycle in which new trees absorb the carbon dioxide released by the trees they are replacing. Over a year that would be equivalent to one return flight to Cyprus.

To see the full list of energy-saving ideas for the home refer to a specialist book.

Move in with someone else

According to Dr Jo Williams of University College, London, people living alone consume 38 per cent more products, 42 per cent more packaging, 55 per cent more electricity and 61 per cent more gas per capita than those in four-person households. They also produce more than half a tonne of extra waste per year. So if you haven't been able to make up your mind about commitment, here's an extra reason to go ahead.

DO YOUR SHOPPING ONLINE

If you buy online you'll not only save the journey to the shops, you'll also make it unnecessary to build any more energy-intensive, customer-friendly supermarkets. Everything could be stored in simple, far more efficient warehouses instead and delivered from there. According to the Department of Transport, shopping trips account for about 20 per cent of car journeys and 12 per cent of the distance driven, so the savings could be considerable.

CAN PERSONAL CARBON OFFSET SCHEMES REALLY MAKE A DIFFERENCE?

Tourism is estimated to be responsible for around five per cent of carbon dioxide emissions. But, of course, that's an average. Some individuals will be responsible for proportionately higher carbon dioxide emissions and some less. Nevertheless, broadly speaking, 95 per cent of your carbon footprint probably comes from all the other things that you do. So cutting your day-to-day carbon footprint is likelier to be easier and more effective than cutting your holiday footprint. That's on the good side.

On the bad side, the average person's holiday footprint may only be five per cent of their total but the picture is going to be completely different for anyone flying long-distance. In their case, day-to-day carbon emissions might have to be reduced to zero – an impossible task – to offset the flight.

In any case, the deepest greens argue that you need to cut your day-to-day carbon emissions by about 90 per cent. And that, even having done so, you *still* wouldn't be allowed to fly. We don't dispute their maths. We simply say it isn't going to happen and that, in the real world, offsetting holiday emissions with savings at home is better than nothing.

Carbon offsetting schemes

In addition to whatever you do at home, you can also contribute to a carbon offset scheme. These specialist companies have the ability to do things you can't. Yes, they plant trees (and you can do the same in your garden) but they're also doing more and more sophisticated things as the industry matures.

Let's take a look at a few of them.

CLIMATE CARE

Climate Care is probably the best-known offset company in the UK and claims to have neutralised one million tonnes of carbon since it was set up 10 years ago. Currently it claims to be offsetting one per cent of UK carbon emissions. That looks optimistic, but even if it only achieves a quarter of that, it's still over a million tonnes a year, equivalent to the carbon footprints of 100,000 people.

What does Climate Care invest in?

Originally, Climate Care did plant a lot of trees but now it's moved on to things like foot-operated water pumps, biogas digesters, wind turbines, hydroelectric generators, fuel-efficient stoves and low-energy light bulbs – all for undeveloped countries. The website includes a quick calculator for your flight. At the time of writing, Climate Care is proposing a donation of £1.99 to offset a return flight from London to Barcelona – which we think is too low.

→ www.climatecare.org

What's wrong with planting trees?

There's nothing wrong with planting trees. On the contrary. We need as many trees as we can get. After all, they're one of nature's ways of scrubbing excess carbon dioxide out of the atmosphere.

But the carbon dioxide crisis is here and now, whereas a tree is in the future. It just isn't going to scrub the gas out soon enough. Your sapling may die or be burned in a forest fire. And even if it survives it may be 50 years or more before it's actually wiped out your carbon emissions. That's way too long. What's more, you can't take credit for all the carbon locked up in the tree because, even if you hadn't planted it, *something* would have grown there anyway. That's how Nature is. To make a real difference, you're going to have to plant a lot more trees than you think. And you're going to have to make sure the scheme has been properly thought through. One offset scheme, for example, has planted non-native pine trees in the upper parts of the tropical Andes. They grow faster than the indigenous species but use so much extra water that they've reduced the flow of some streams by 70 per cent.

What's more, Stefan Gössling of Lund University, Sweden, has calculated that around 30,000 square kilometres would have to be planted with trees every year in order to offset current emissions from air travel. That's an area the size of Belgium. Ireland would be good for only a couple of years and the whole of the UK would take care of less than a decade's worth of emissions.

So, yes, it would be wonderful if we could plant our way out of the carbon dioxide crisis and trees should be part of any offset portfolio. But you can't just pay for a tree or two and think you've done enough. You haven't.

CLEAR

Clear takes a very different approach. Your money will be used exclusively to buy what are known as Certified Emissions Reductions (CERs) under the now famous Kyoto Protocol. As usual, the bureaucrats have dressed the whole thing up in such convoluted language as to make the whole subject almost unintelligible. But in essence it amounts to this. You'll be guaranteed that your money will translate directly into a reduction in carbon dioxide emissions via projects registered with what's called the Clean Development Mechanism (CDM). In other words, 'it's official'. By November 2007, 828 projects had been registered by the CDM Executive Board and were calculated to have decreased greenhouse gas emissions by the equivalent of around 171 million tonnes of carbon dioxide a year. (We say 'equivalent' because sometimes the reduction involves a different but similarly dangerous pollutant.) A further 2,600 projects are in the pipeline, which are projected to decrease emissions by the equivalent of about two-and-a-half billion tonnes of carbon dioxide up to 2012. To put that into perspective, the EU-15 countries produce the equivalent of about four billion tonnes of carbon dioxide a year. So, although still not enough, the impact could be significant.

But it has to be pointed out that CERs are not beyond criticism. Sceptics say many of the projects would have happened anyway and it's not a genuine saving. And some companies have succeeded in reducing emissions for far less money than they've been paid under the scheme, thus making substantial profits out of it. But these teething troubles are gradually being overcome.

→ www.clear-offset.com

CARBON PASSSPORT

This organisation invests your money in practical projects such as collecting landfill gas and subsidising renewable energy so that developing countries can afford it. The calculator on its site suggests £5.28 as the correct offset for a return flight from Stansted to Barcelona (Girona) – almost three times the Climate Care figure (and probably far more realistic).

→ www.carbonpassport.com

CARBON NEUTRAL

Energy-saving gifts and offset certificates as presents are the special feature of this website. Here you can buy low-energy light bulbs, for example, recycled glassware and wormeries. But don't forget, it's better not to give any present than an unwanted present, even if it was produced using low-energy methods. Carbon offset packages include clean technology, forestry and projects in developing countries.

→ www.carbonneutral.com

Tour operators and carbon offset

If you're thinking of booking a holiday with a tour operator check to see if carbon offsetting is available. Some travel companies include carbon offsetting in the price of the holiday. Others include an offset charge in the invoice but say you don't have to pay it if you don't want to. Climate Care says take-up can be as high as 70 per cent under this scheme. Others provide the opportunity to opt in to a scheme but, in these cases, take-up is only around 15 per cent.

Whatever method your tour operator uses, pay the money and be happy to do it. It's far from perfect. But then, what is?

If your tour operator doesn't have a carbon offset scheme then ask why not. The big tour operators believe most people don't care. Let's show them that we do.

So who's right?

We look at it this way. You shouldn't use the concept of carbon offset as a way of carrying on wasting energy. But if you have little choice but to consume energy then surely it's better to donate money to a good cause than to do nothing at all? Well, isn't it? Don't fly to New York to go shopping when you can buy everything you need in your local high street or on the internet. Don't fly to the Caribbean for a beach holiday when you can have a perfectly good beach holiday closer to home. But if you want to see the Sistine Chapel and the Colosseum and you drive to

Rome in an energy efficient car full of passengers, then why is it wrong to plant trees or provide third-world countries with low-energy light bulbs?

Here's a summary of the arguments to help you make up your own mind.

Against Carbon Offset For Travel	**For Carbon Offset For Travel**
Offsetting isn't the answer; you just shouldn't travel in the first place.	There's no reason to single out travel; it isn't immoral. On average, 95 per cent of your carbon footprint is likely to come from running your home, going to work and the things you buy.
A lot of the money is spent on things that would have been done anyway.	That's like saying you shouldn't give money to charity because somebody else will probably sort the problem out.
Providing poor people in the Third World with things like foot-operated water pumps so people in the Rich World can continue to burn fossil fuels is just a form of exploitation.	How can helping people in the Third World be wrong? Would you prefer your money to be spent on, say, energy-saving light bulbs for people in the West?
Carbon offset encourages people to think they can travel without contributing to global warming when, in reality, they can't.	The majority of people undoubtedly realise that carbon offsetting is only a contribution to the solution, not the whole solution; but it's better than nothing.

Chapter 4

When you get there

FACTS YOU NEED TO KNOW

- Millions of hotel beds are unused most of the year – a terrible waste of resources; home swaps, private rooms and camping are the green alternatives.
- Tourism effectively transfers water from southern to northern Europe and from the Caribbean to North America.
- The bushmeat trade is decimating wildlife on an unsustainable scale.
- Animal viewing holidays help protect wildlife.

So you've got to your holiday destination. But now what? Should you stay in a high-rise hotel and save the maximum amount of land? Or should you stay in a villa, tucked inconspicuously into a hillside? Should you eat the local food, or should you leave it for the locals? Should you go to see wildlife or is it better to lie on the beach and leave the animals in peace?

These are the sort of questions that don't always have clear cut answers. But we'll do our best.

Accommodation

A little ball of dried weeds blows along the dusty road. The shops are closed and boarded up. The hotel windows are shuttered. The car parks are empty. No one walks along the streets. A sign hangs from one corner, creaking in the wind.

But this isn't some Wild West ghost town. It's the Costa Brava in January. And the story is the same all along the coast. Around the Mediterranean alone there are enough tourist beds to close many of the world's refugee camps – some 10 million, in fact – and almost all of them are empty half the year.

It's a terrible, un-green waste. But are there really any alternatives?

HOME SWAPS

The home swap is the perfect answer to green holiday accommodation. No habitat is destroyed. No resources are wasted. No mixer-loads of concrete need to be poured. No beds are left empty. Nothing is specially built for tourists. You go to stay in someone else's house and they come and stay in your house. Simple. Logical.

But how do you find someone to swap with? In fact, in the age of the internet it's easy. There are several online agencies and all you have to do is register with one of them. And it's not expensive – think in terms of around £50/£100 a year. Obviously you won't be allowed to access details of potential swappers' properties until you've paid your money, but most sites will allow you to see a limited selection to get an idea.

When you upload your own details and photographs, be accurate. There's no point in saying you've got three double bedrooms if, in reality, you've only got two plus a cupboard. That will only make for bad feeling when, in reality, the whole process relies on trust.

It's also important to be flexible. It may well be that you're hoping for a place in, say, Marbella, but you might have to wait a long time until something so specific comes up. Be willing to go for other similar Spanish resorts. Better still, just see what's on offer on the site and choose whatever appeals the most.

Obviously you won't want to let strangers into your home without very careful thought. So it's a good idea to make sure they *aren't* strangers.

At least, not by the time you actually do the swap. In other words, register with an agency well before you actually want to travel. A year is about right. That gives you plenty of time to find a fellow swapper and to get to know one another. If everything goes well, maybe you can do the same the following year without the hassle.

If you have children they may be less enthusiastic about the idea than you are. They probably won't like the idea of other children invading their private spaces so be sure to consult them fully before making a commitment.

Almost certainly someone will be getting a better house and someone will be getting a less comfortable house. If you're on the losing end of the deal, just shrug it off. It's not going to be possible to find two identical houses to exchange. The important thing is that you have a great holiday – without harming the environment.

Here are some sites to take a look at:

➜ www.homelink.org
➜ www.homebase-hols.com
➜ www.homeexchange.com
➜ www.another-home.com
➜ www.swaphouse.org/eng

TIPS

- Register with an agency well in advance of your intended trip.
- Be honest about your property and the surrounding area – you don't want recriminations later.
- Take your time getting to know the other people – speak on the telephone and exchange photographs of yourselves.
- Make sure you have someone (a neighbour, a relative, a friend) who can keep an eye on things and deal with any emergencies.
- Lock away personal and valuable items – if possible, close off one room and store your 'treasures' in it, together with clothing and any other items you have to move to make space for the arrivals.

- Disconnect the telephones – or make a clear agreement about how the telephone bills (and incoming calls) are to be dealt with. If the telephones are disconnected, both parties should be equipped with mobile phones.
- Leave your own house as clean, tidy and welcoming as possible – most people will respond by taking far more care of your property if you do.

Debbie's experience

Our home swap took an awfully long time to organise. We registered with an agency at the beginning of July, hoping to go away in August but, in fact, it took until the *following* August. Basically, they wanted to know a lot about us before they'd let us into their house, and we were the same. We were feeling one another out for several months before we all decided to go ahead. We e-mailed pictures of ourselves and spoke on the phone several times. There was also the problem of co-ordination. In the end, they came and spent one night in a hotel close to our home. We showed them around, exchanged spare sets of keys, and set off for their house the next day. On the return, three weeks later, we travelled simultaneously so we didn't meet again. I'd steeled myself for breakages and so on but, apart from a glass or two, there was no physical damage. I'd definitely do it again for a long stay.

Private rooms

When you think about it, most homes also have a lot of wasted space. For a start, almost all of us have a 'spare room' just in case someone comes to stay. But they very rarely do. And if you're over, say, 50 you've quite likely kept the kids' bedrooms exactly as they were. Only the kids grew up, left home years ago and only come to stay every other Christmas. All over the world that adds up to an awful lot of unused bedrooms.

But some people don't leave them unused. They make them available to tourists. Tourists just like you. Years ago, bed and breakfast in private houses was the cheap option for holidaymakers who just couldn't afford to go away otherwise. Today, private rooms are still great value, but that's not the main reason for choosing them. Here are some of the advantages:

- You get to meet local people.
- You have a marvellous source of local knowledge.
- You have a wonderful opportunity to practise the language.
- You can stay away from 'touristy' developments.

Obviously it's impossible to generalise. Everyone's home is different. But with the rise in living standards in the developed world in the past few years, many private rooms are the equal of what's on offer in top class hotels and include private facilities.

If you're on a touring holiday you can just watch out for the signs along the road. They may say something like *Zimmer Frei* or, indeed, *Bed & Breakfast* because that's now an international phrase. But if you want to book something in advance it's best to look on the internet or get in touch with the tourist office.

The tourist offices in many countries will run an accreditation scheme for private rooms, grading them according to size, comfort, cleanliness and facilities. Some also award eco-labels. Most big cities and tourist areas will have a thriving bed-and-breakfast sector, but there are certain countries where, by tradition, you'll seldom have any problem finding a comfortable private room. They include Scandinavia, Germany, Austria, Switzerland and, indeed, the UK.

It would be impossible to list all the relevant websites. The best thing is to enter the name of your destination in your search engine, together with, say, *bed and breakfast* and see what comes up. Meanwhile, here are a few to get you started:

➜ www.bnb.choices.com – to date covers 62 countries all over the world
➜ www.bedandbreakfastnationwide.com – United Kingdom
➜ www.britainexpress.com – United Kingdom
➜ www.londonhometohome.com – London
➜ www.londonbb.com – London
➜ www.parisbandb.com – Paris

→ www.bed-and-breakfast.de/index_en.asp – Germany
→ www.fewomat.de/index.html – Germany and Europe
→ www.zimmer-im-web.de/indexe.htm – Germany and Europe
→ www.go-lodging.com – USA and Canada
→ www.bbdk.dk – Denmark plus Scandinavia, Iceland and Baltic countries
→ www.bba.nu – Stockholm

Meet the locals

If you've never stayed in private homes before, Denmark is a great country to get started. There are rooms all over the country and marvellous get-to-know-the-locals schemes such as *Meet The Danes* [*see below*]. And very nice they are, too.

In Copenhagen you can go to the Tourist Information Centre:

→ Vesterbrogade 4A, 1620 Copenhagen V.

Otherwise, you can book in advance on a number of websites:

→ www.visitcopenhagen.dk – the official tourist office website
→ www.meetthedanes.dk – private rooms and homedinners in Copenhagen
→ www.useit.dk – mostly for young people/students in Copenhagen
→ www.cosydenmark.dk – hosts all over Denmark
→ www.bedandbreakfast.dk – in the countryside as well as towns
→ www.bed-breakfast-fyn.dk – specialists for Funen
→ www.bondegaardsferie.dk – specialists in rural properties

Camping and caravanning

A tent is, undeniably, a pretty ecological form of accommodation. A certain amount of material and energy goes into its construction but, compared with a hotel room, it's hardly worth mentioning. The campsite is a different story. At the very minimum it will have toilets and showers but quite probably a restaurant, a shop, tennis courts, a swimming pool and possibly more. And as camping is a one-storey affair it will probably spread out over quite a few acres. Nevertheless, canvas beats a hotel.

Camping wild is even better. In fact, it's *real* camping. You hear the owls at night, not someone snoring in another tent just three paces away from yours. Of course, camping wild is forbidden in a lot of places. Mostly because they're not wild enough. Locals don't want to see a row of tents round the edge of the local common. But get truly away from civilisation and nobody is going to say anything.

In Britain there is no right to camp and you'll have to seek the permission of the landowner. In upland areas, however, wild camping is usually tolerated. In Scotland the position is more liberal and there *is* a statutory right to camp in appropriate circumstances – you can see the details at www.outdooraccess-scotland.com. In remote areas of the Alps, Pyrenees and Scandinavia there's no problem at all.

We haven't got space here to go into all the techniques of wilderness camping. For that you need a specialist book. But to ensure you save the environment rather than damage it, here are a few pointers:

- Try to pitch your tent on bare earth so as not to damage the vegetation; if you have no choice but to pitch on grass move the tent to a new position every day.
- Don't camp too near water – it's tempting but you may scare away animals that depend on it.
- Don't engineer the site to make it more comfortable – if you have to move rocks and cut branches then find a more suitable site.
- Never light a fire if there's any fire risk; if you do light a fire be sure to disperse the ash next morning and leave the site as unblemished as possible.
- Attend to bodily functions well away from water – for solids, dig a "cat hole" and fill it in afterwards. Paper should be burned, provided there's no fire risk; if there is a risk, seal the paper in a plastic bag and take it home.
- Take all rubbish with you.
- Leave the site so that no one would ever know you'd been there.

Caravans and motor homes are a slightly different story. A lot of material and energy goes into their creation which, from the environmental viewpoint, isn't really justified by one week's use a year. If that's your situation, consider renting rather than buying.

Farm stays

Farmers nowadays have to cultivate a lot more than wheat if they want to make a decent income. Plenty are now cultivating tourists. Essentially, this is a variation on bed and breakfast and from the environmental point of view it's a good one. Farms tend to have a lot of outbuildings which often aren't put to very good use, so converting part of a barn into a bedroom makes good sense. And if your food comes from the farm there couldn't be any finer carbon-saving arrangement.

If you don't want to be 'in the middle of nowhere' a farm stay could still be for you. Most farms aren't actually so very far from civilisation and quite a few are close to beaches. Meanwhile, any members of the party who want tranquillity, animals and country walks are going to be very happy.

A good place to start tracking farms down is www.agrisport.com which provides links to farm stays all over the world. And here are some sites for specific countries:

- ➜ www.farmstayuk.co.uk Tel. 024 7669 6909 UK
- ➜ www.farmbreaks.org.uk UK
- ➜ www.irishfarmholidays.com Tel. 00 353 61 400 700 Ireland
- ➜ www.agriturismo.com Italy
- ➜ www.bretagnealaferme.com Brittany, France
- ➜ www.ecoturismorural.com Spain
- ➜ www.bauernhof-ferien.ch Tel. 00 31 329 66 33 Switzerland
- ➜ www.farmholidays.com Austria
- ➜ www.tiscover.at Austria
- ➜ www.farmholidays.de Germany

Self-catering

It's impossible to generalise about self-catering accommodation because every case is different. Some self-catering units are basically hotel suites with extra facilities. Others are high-rise holiday apartment blocks. And still others are traditional, local houses that have been renovated for tourists. These are the ones to go for.

When you go to stay in, say, an old, restored Catalan *mas* you're achieving several things. You're:

- Helping local people earn a living in the countryside.
- Reversing the earlier decline in country properties.
- Making use of assets that were going to waste.
- Having a very nice time.

→ www.click4holidayhomes.com
→ www.countrycottagesonline.net
→ www.holidaylets.net
→ www.holidaylettings.co.uk
→ www.holidayrentals.org
→ www.holiday-rentals.co.uk
→ www.ownersdirect.co.uk
→ www.PerfectPlaces.com
→ www.self-catering-breaks.com

Hotels

Today's giant hotels concentrate a lot of people into a small ground area. So you could argue they're good for the environment. On the other hand, high-rise development has ruined huge lengths of coastline all over the world. Realistically, it's a question of what's appropriate. In London, New York, Hong Kong and so on, it's difficult to argue that tall buildings are aesthetically damaging. In fact, many people think the opposite. But it should be remembered that cement produces somewhere between five and 10 per cent of man-made carbon dioxide emissions world wide. So if we can cut back on wasteful construction we can save a lot of carbon dioxide. There's certainly no argument that along the coast and in the countryside all development should, ideally, blend into the landscape.

Here are some websites that will help you find small hotels:

→ www.seclusionholidays.co.uk – UK
→ www.silencehotel.com – Europe Tel. 00 33 1 44 49 90 00
→ www.secretplaces.com Tel. 00 35 121 464 7430
→ www.secretdestinations.com Tel. 0845 612 9000

Ecological accommodation – is it really green?

Increasing numbers of developments are now billing themselves as 'ecological'. It's good marketing. But are they really? Too often it simply means a nice view of the countryside.

At the end of the chapter we'll be giving you some questions you can ask your hotel or, indeed, any company involved in your holiday. But you can tell a lot from first impressions. For example, are there immaculate lawns that need watering? Are there swimming pools, even though the accommodation is close to the sea? Are there lots of unnecessary lights on? Are they ordinary (that is, not low-energy) lights? Are the rooms uncomfortably hot? Or over air-conditioned? Does everything in your room stay on when you go out? Are the windows single-glazed? Is bottled water on sale in the bar (wasteful, except where the tap water is unsafe)? And take a look round the back, behind the kitchens. Is the waste all thrown together in unsorted bins? If the answers to these sorts of questions are generally 'yes' then the management just isn't serious.

Water – part 1

A survey on the tropical island of Zanzibar found that tourists in guesthouses used 100 litres of water a day whereas those in the luxury hotels used 2,000 litres a day (directly and indirectly). The big hotels used far more on things like swimming pools, poolside showers, laundry and irrigating the grounds. A lot of it is crazy. If a hotel is close to the beach, for example, it doesn't need a swimming pool. Swimming pools use a lot of energy in their construction and operation, so it's best to stick with what nature provides.

Stefan Gössling of Lund University, Sweden has calculated that international tourism is, in effect, a sort of water transfer from one part of the planet to another. According to his calculations, northern Europe saves 140 billion litres of water a year as a result of its citizens going on holiday in southern Europe and other places, while North America saves 43 billion litres. Meanwhile, southern Europe 'loses' over 200 billion litres and the Caribbean 'loses' over 20 billion litres. Neither region can easily afford so much water.

Water – part 2

It was surprising how long it took before we realised we were in a very dangerous place. At first the water had looked clear and the little rocky pinnacle, sticking up about a metre above the water, had promised good snorkelling. And, indeed, there was quite a lot of activity in the water. But not what we expected. As we got our bearings, looked about and focused on the shapes all around us we suddenly comprehended they weren't fish. We had swum down onto the outfall of a sewer. Two people never rose more quickly to the surface. We kind of breached.

And that was in the Mediterranean only a few years ago. For all we know, the situation in that particular spot hasn't changed. The fact is that, despite a huge investment, many Mediterranean coastal areas still can't cope with the high-season sewage problem.

Food

The first very clear thing to say about food on holiday is this: Never be tempted to eat bushmeat – that's to say, wild animals that have been trapped or shot. And, of course, don't eat it at home either. The bushmeat business, as we saw in the *Introduction*, is destroying much of the planet's wildlife.

To recap, it's estimated that one million tonnes of bushmeat is taken from Central Africa each year. That's equivalent to nine billion quarter-pound hamburgers. It's just not sustainable. So don't try bushmeat, even out of curiosity.

But what about wild animal 'farming'? Isn't that the way forward? No, it isn't. In the first place, farming wild animals is generally far less profitable than farming domestic animals. People who can't afford the high prices will still hunt and, meanwhile, the very existence of the farms will effectively 'launder' bushmeat. In other words, anybody caught with bushmeat will be able to say they obtained it from a farm. Furthermore, the high concentration of animals on the farms could easily become a breeding ground for diseases that could spread into the wild populations.

The only answer to the bushmeat problem is to provide alternatives for those driven by poverty, and to change attitudes in those with money. The same applies to hunting wild creatures everywhere in the world, of course. It isn't just a question of Africa or Indonesia. Wild deer from the Scottish highlands and grouse from English moors are just as much bushmeat as a kudu or a monkey.

What's often overlooked is that hunters disturb and kill far more than the species they're after. When men with guns can't find the target species a proportion will shoot anything else they see. We've been to bird rescue centres and seen, for example, eagle owls and herons that have been shot. With help, these large birds can sometimes recover. Smaller protected species are simply blasted to death.

What's more, hunters manage the environment so as to have the maximum amount of 'sport' and that often means trapping other species. You won't see many birds of prey on a typical grouse moor, for example. And whilst protesting that they're ridding the countryside of 'pests' they often feed, and even breed, those very same target species so as to be able to shoot them later.

We live in a part of Spain, for example, where wild boar hunting is a tradition. The local farmers complain that the wild boar destroy their crops. And yet those very same men *feed* the wild boar so as to keep them in the area and increase their numbers. What's more, they've deliberately allowed them to interbreed with domestic pigs so as to increase the size and frequency of litters. The net result of hunting that particular species is therefore an *increase* in wild boar numbers, to the detriment of certain other species, especially ground-nesting birds and tortoises, as well as the farmers' own crops. It's madness. People get killed every year in hunting accidents. And, on hunting days, large areas are just too dangerous for ramblers and nature lovers to venture into them. This is the kind of thing that happens when people believe they can do whatever they like with wild animals and wilderness.

Well, those wild animals don't belong to the hunters. If they 'belong' to anybody they just as much belong to you and me. And *we* don't want them killed. These relatively few people have no right to deprive the rest of us – the majority – of the pleasure of seeing wild creatures. In a very real way, our human rights are being infringed.

You're probably entirely in agreement with us so far. But what if we tell you one of the biggest things you can do to offset your travel emissions is, well, to give up eating meat *entirely*. Why?

- According to the US Environmental Protection Agency, farm animals produce almost a quarter of global emissions of methane. Methane, you'll remember, is a greenhouse gas that traps many times more heat than carbon dioxide. And one cow alone produces 350 litres of methane per day. As to carbon dioxide itself, a cow's output is around 1,500 litres per day. On top of that, huge areas of forest – the planet's mechanism for scrubbing carbon dioxide out of the atmosphere – are being cleared for livestock, making the problem even worse.

And here are a few more environmental reasons for giving up meat:

- The amount of land needed to support the typical western diet is about one-and-a-quarter acres. But on a mainly vegetarian diet it's about half an acre. On a completely vegan diet it's something like a fifth of an acre.

- Domestic animals generally don't increase the amount of protein available for humans to eat. On the contrary they reduce it, because the animals themselves have to eat. From that point of view, beef is the worst. It takes something like ten pounds of vegetable protein to produce one pound of beef protein, a catastrophic loss.

- The energy required to produce food for a meat eater is more than three times greater than for a vegan.

- The water required to make a day's food for a meat eater is 13 times more than for a vegan.

There are health arguments, too, but that's not the subject of this book. When you realise that there are more farm animals on the planet than there are people you begin to understand the scale of the situation. If nobody ate meat the health of the planet would be a lot, lot better than it is now.

Should you watch wildlife?

Wildlife holidays might seem to be the very essence of green tourism. But are they really? Quite a few greens say they're not. In the first place, you have to get to where the animals are, producing quite a lot of pollution in the process. Then you need somewhere to stay. Which means building hotels in what was previously a wilderness. Then you need to be close enough to see. Which means constructing roads and tracks, on which, inevitably, some animals will get run down. Finally, the viewing itself is likely to disturb the animals.

Those are some of the arguments for leaving the animals in peace. It's persuasive but we think there's a flaw. It's this. The animals *aren't* going to be left in peace. Someone is going to want to cut the trees for timber. Someone else is going to want to use that land for agriculture. Someone else will want open cast mining. And so on.

Unfortunately, we live in a world dominated by money. You may not like it. But that's how it is. And you're not going to change it.

So either we work with it. Or we fail. In other words, if live elephants generate less money than agriculture would on the same land it's goodbye elephants. If live dolphins are worth less than tinned tuna it's goodbye dolphins. If live whales are worth less than whale meat it's goodbye whales. And so on.

And those are all high-profile animals. If they can't be saved, what hope is there for, say, the red-headed vulture or seaweed?

One in eight species of birds, a quarter of the world's mammals, one third of all amphibians, and almost three-quarters of the plants that have so far been assessed by the International Union for the Conservation of Nature (IUCN) are threatened with extinction. That's disastrous. And as if emphasis were needed, the Yangtze river dolphin actually did become extinct in 2007. That's a tragedy.

Well-meaning people, paid or working for free, can only do so much. Even in the face of environmental disaster only one thing speaks. And you already know what that is.

So *how* can you give monetary value to wildlife? The answer is, by going on holiday and paying to see it. And we all have to be willing to pay a lot. Which is hard. Not just because spending money is a painful thing to do but because it seems to go against our natural right to travel freely in wilderness areas.

Why should we have to pay for something that used to be taken for granted? The answer is simple. Because if we don't, it just won't exist any more. There won't be any wilderness to explore, except in the most difficult regions, and there won't be any wild animals to see.

Many environmentalists are now arguing against travel, and especially air travel, because of the pollution it causes. And, of course, they're right about that. But if nobody travels to see wild animals many species will become extinct. And very quickly. For some animals it's hard to tell which would come first – extinction by global warming or extinction by economic circumstance.

What's the alternative? You could simply pay the money and *not* go. Kenyans could pay more taxes to preserve their parks. Or everyone in the world could make a contribution. And to all the other worthwhile projects, too. There's a strong argument to be made for a world wide tax to preserve the environment for everyone's benefit. But it just isn't going to happen any time soon.

Of course we're aware that during a period of unprecedented tourism growth, the problem has only grown worse. In just one year, the International Union for the Conservation of Nature (IUCN) added nearly 200 more species to its famous Red List – the catalogue of threatened wildlife. In 2007 the total stood at 16,306 species. But we think the situation would be graver still without tourism.

Yes, spiritual values count. And the fact that you're reading this book suggests they mean a lot to you. But not everybody feels the same way. Especially not those who have little to eat. They'll go along with wildlife conservation if it can put some money in their pockets. That's why, in Chapter 11, we'll be describing some long-distance wildlife holidays. It would be better if you went by train, coach or truck but we think there are occasions flying is permissible. These are such a case.

Some questions to ask before you book

- What is your company doing to become carbon neutral?
- What are you doing to reduce energy needs?
- Are you using clean energy?
- Is a proportion of my money being used for carbon offsetting?
- Is a proportion of my money going to support environmental projects?
- What proportion of my money will remain in the destination?
- What's your policy on employing local people?
- How much (building materials, furniture, furnishings, food) is sourced locally?
- Are your vehicles (planes, trains, coaches) the latest, fuel-efficient designs?
- Have you installed any solar voltaic panels or solar water heating?
- What water conservation measures have you introduced?
- Do you have swimming pools, even though you're near the beach?
- Are sewerage systems able to cope with holiday peaks?

Chapter 5

The truculent traveller

FACTS YOU NEED TO KNOW

- The travel industry is the largest in the world, which means tourists wield enormous power.

When Charles Darwin landed in the Galapagos Islands in September 1835 he was astonished as much as anything by what in his journal he called 'the tameness of the birds'.

He describes how he lay on the ground holding a pitcher of water and a mocking-bird alighted on the rim of it. All the birds, even the hawks, came so close it was possible to touch them.

What a contrast with the world today where we need powerful binoculars to get a good view! And did Charles Darwin appreciate the enormous privilege? Did he feed the birds? Did he delight in having them perch on his hands? Did he stroke them? Not a bit of it. He killed them. And so did all the other members of the crew of the *Beagle*. They caught them with their bare hands, they swatted them with sticks, they shot them. They had Paradise in their grasp and they all – even a scientist like Darwin – destroyed it.

In our more enlightened times, of course, people wouldn't do that. Would they? Well, yes they would. And do. In the Mediterranean region alone, hundreds of millions of migrating birds are shot and trapped every year. No, that's not a misprint. *Hundreds of millions.*

It's a tragedy in so many ways. But a tragedy that you, as a tourist, could help put an end to.

Tourists have enormous power. Just think about it. This is the biggest industry in the world. Which means that green travellers have the power to do a lot of good.

Just suppose, for example, that tourists *en masse* were to say that for a year:

- They're boycotting Spanish holidays because of bullfighting, or
- They're boycotting French holidays because of unsustainable bird hunting, or
- They're boycotting Japanese goods because of whaling.

The impact would be enormous. But boycotts have to be organised on a big scale if they're to have any effect. In the absence of a boycott what can you do in the meantime as an individual traveller?

The message is this. Wherever you go on holiday – and wherever you don't – *protest*!

Here are some environmental issues in popular holiday destinations:

Bullfighting

Bullfighting goes on in most of Spain (but no longer in Catalonia), the south of France, Portugal, Colombia, Ecuador, Mexico, Peru and Venezuela. That doesn't mean, of course, that everybody in those countries supports it. Obviously, you wouldn't go to a bullfight when you're there but you can do more. If other holidaymakers in your hotel are planning to go, try to persuade them not to. If your tour operator offers a bullfighting trip then you should complain. If you discover that a bullfight is coming to your resort go to the tourist office and make your views known. But never be aggressive. That just makes people more entrenched in their positions and, anyway, many people will agree with you. When the silent majority gets a little bit more noisy things may change. If you'd like to do more, contact the League Against Cruel Sports:

→ **www.league.org.uk** Tel. **0845 330 8486**

Whaling

Do you remember that campaign from years ago? *Stop the bloody whaling.* Well, some did but a few didn't. And those who didn't are even now agitating to have restrictions on whaling lifted. At the time of writing, a Japanese fleet has just set sail in search of minke and humpback whales. Yet this *is* an incredibly bloody business. We're talking about highly intelligent mammals, after all.

The countries that are the main culprits are Norway, Iceland and Japan.

Tourists can help save the whales in two ways. Firstly, by paying to see live whales in the wild. That's to say, whale watching trips. The aim is to make live whales more valuable than dead ones. Secondly by boycotting anything to do with whaling. If you're in a restaurant where whale meat is served, for example, you should leave after explaining why you're doing so. You could go further and boycott products from the countries concerned – again explaining why. But should you boycott the country itself as a tourist? There's a dilemma about this. Many environmentalists think you should refuse to visit whaling countries. We take the opposite view. We believe you should go to the country and *pay* for whale watching. Of course, make your anti-whaling views known in the appropriate quarters. But it's by giving significant economic value to living whales that whaling is most likely to be defeated for all time.

For more information about the current state of whaling see www.greenpeace.org.

Animals in captivity

Many greens are totally against ever keeping animals in captivity. But there are also powerful conservation arguments in favour of zoos and wildlife parks. We take the case by case approach. If there had never been any captive breeding programmes – which some zoos are very good at – then we would have lost more species than we have so far. What's more, zoos educate people and, indeed, sensitise them. It's much more likely someone will care about the fate of, say, pandas once they've actually seen a panda.

But we can all agree there's no justification whatsoever for keeping animals in bad conditions as a way of making money. So if, on holiday, you visit a zoo where animals are mistreated then you should try to do something about it. In particular you should contact the Born Free Foundation which campaigns on behalf of animals in captivity:

➜ www.bornfree.org.uk

Hunting

Hunters argue that they give monetary value to wildlife and therefore help preserve it. The argument may stand up for some species but, in general, hunters have decimated wildlife all over the globe and been directly responsible for the local or total extinction of several species.

There's absolutely no merit in the argument when it comes to the hunting of wild birds, for example, which goes on everywhere but is particularly serious around the Mediterranean. As we've already said, hundreds of millions of birds are shot or trapped every year in Spain, France, Italy, Greece, Syria, Lebanon, Tunisia, Cyprus and Malta. Some of this hunting is legal and some is illegal but, whichever kind it is, the result is fewer birds in those countries and fewer birds back in Britain.

Hunting is a lobby that punches far above its weight. Yes, it involves large numbers of people and, yes, they spend quite a lot of money. But, nevertheless, they're very much in the minority. In France, for example, there are 1.3 million people who hunt as against about 60 million who don't. In Spain 980,000 people hunt and about 40 million don't. The hunters make a lot of noise and wield a lot of power but if the people who were opposed to hunting would only get together in the same way then bird hunting would become much more restricted if not halted altogether.

Part of the solution is to match the hunters' economic power with the environmentalists' economic power. Hunters are prepared to pay to shoot and we have to be prepared to pay to see wildlife. When you visit a protected area don't begrudge any payment you're asked to make. And if you're not asked, make a donation anyway. It's important that protection pays.

What else can you do? If you find yourself in a restaurant where, say, ortolan buntings are on the menu then leave, after politely explaining your reasons.

If you'd like to do more you could write to the appropriate ministry in the country concerned. Here are the addresses for the two most popular holiday destinations:

DG Nature Conservation,
Ministry of Environment,
C/Gran Via de San Francisco 4,
Madrid,
SPAIN
E-28071

The French Ministry of Environment,
57, rue de Varenne,
Paris,
FRANCE
F-75700

The following organisations campaign against illegal bird hunting as well as for tighter restrictions:

- The Royal Society For The Protection Of Birds www.rspb.org
- The Sustainable Hunting Project, BirdLife International www.birdlife.org

If you'd like to join the campaign against *all* hunting, contact the League Against Cruel Sports [*see Bullfighting above*].

Tip

If you want to get involved it's always best to work with local activists to give support in a low-profile sort of way. Nobody likes it when 'foreigners' start telling them what they should or shouldn't do. It can be counterproductive. So get in touch with the local people who are already working on the ground and find out how you can back them up.

However, as a tourist there's no reason you shouldn't raise your concerns directly with the tourism authorities. That's different. Let them know that you found, for example, bird hunting distressing and, indeed, potentially dangerous.

Chapter 6

Do you really need to go?

FACTS YOU NEED TO KNOW

- Technology can make many journeys unnecessary – provided we all adapt to the new culture.

Supposing you could talk to business associates, or even family members, thousands of miles away, as if they were in the room with you? Would you still feel the need to make that trip?

Before you answer that, we'd better set the scene in a little more detail. You'd be sitting at one end of a table and they'd be at the other, just as in a conference room. Or rather, they'd appear to be sitting at the other, courtesy of high-definition life-size screens. You'd make eye contact when you spoke. When they replied, the sound would appear to come from their mouths, not a speaker off to the side. And there'd be no time delay.

If you've already had some experience of videoconferencing it probably wasn't like that at all. The image was probably jerky, the voice probably wasn't synchronised, and because you were all focusing on the screen rather than the lens, there wouldn't have been the impression of eye contact. But the scene we've described isn't the technology of the future. It's now. It's the world of 'telepresence', the latest, hi-tech version of videoconferencing.

There's little doubt it will reduce the growth of business travel. For the moment telepresence is enormously expensive. It works out at about £200,000 per installation and only big companies can afford it. But like all things electronic, the price will come down significantly in the near future. And, of course, you can always hire [*see below*].

The advantages are fairly obvious. Time and the environment. If you don't fly from London to New York for a one-hour meeting you'll save, at the very least, 24 hours. Realistically, more like two days. You'll also save something like four tonnes of carbon dioxide. Which is enormous. It's the equivalent of roughly three to four months of emissions, if you have a typical UK carbon footprint.

Here's another scenario. You're recruiting in several different countries. Normally you'd have to pay a fortune for travel and hotels. But, in fact, you don't have to go there at all and the candidates don't have to come to you. All you have to do is hire a videoconferencing suite in each country, settle down in yours, and conduct the interviews remotely.

Or what about this? A client on the opposite side of the world is interested in your product but would like to see various functions demonstrated. Normally you'd despatch a salesman. But why? All you need do is hire a videoconferencing room somewhere convenient for your client, invite your client along, and demonstrate your product without leaving the UK.

Of course, it requires the adoption of a different culture. We have to learn to put less emphasis on physical presence. We have to learn to assess people – and trust people – by a video image. And we will.

Hiring facilities isn't expensive. In London and other capitals think in terms of around £100 for the first hour during normal office hours, £150 in the evenings and £200 at weekends, with a discount for additional hours. Provincial prices are lower.

Here are some companies that can help:

➜ **www.eyenetwork.com** Tel. **0845 355 0290**
info@eyenetwork.com – essentially a videoconference booking agency representing some 3,000 studios worldwide.

- → www.londonvideoconferencing.co.uk Tel. 0207 236 8686 info@bridgesvc.com
- → www.video-conference-london.co.uk Tel. 0207 255 2557
- → www.eoffice.net Tel. 0870 888 88 88 eo@office.net – facilities in London, Birmingham, Manchester and Bristol

Webcams

In 1991 some people in the computer science department of Cambridge University pointed a camera at their coffee pot and hooked it up to the world wide web. Thus the first webcam was born and the coffee pot was its first star, appearing daily for 10 years.

It may seem unbelievable but there are now literally millions of similar webcams set up all over the world. They point at famous buildings, busy streets, quiet streets, mountains, beaches, ski resorts, dance floors, people's dogs and much, much more. In other words, on your computer, you can connect to remote cameras in a mind-boggling variety of places to see what's going on. Right now. And, when you think about it, that can save you a lot of journeys.

Let's say, for example, you're thinking of moving to a new area 100 miles away. You've heard it's nice but you don't really know. Normally you'd get in the car and go see for yourself. But there's another way.

All you have to do is sit in front of your computer, access an index of webcams and click on the ones in the target area. With luck there may be several and you'll be able to make a judgement without leaving home.

Or maybe you're thinking of going away for the weekend. You've seen a general weather forecast, but can you really trust it? You don't have to. Go to the relevant webcam and what do you see? Rain. So you stay at home.

Or you were going to travel to a minor sporting event. You'd be happy to watch it on TV but it's too small to be covered. However, it might not be too small for a webcam. And, indeed, there it is on the index. So you settle into an armchair and save yourself a journey.

And how exactly do you do this? It's very simple. Once connected to the internet just put the destination or event into your search engine together with the word 'webcam' and see what comes up. Alternatively, you can go to a webcam index to see what's listed. Here are a few to get you started:

→ www.earthcam.com
→ www.camcentral.com
→ www.webcam-index.com
→ www.webcamsearch.com
→ www.webcamworld.com

You can also see satellite images of most places on Earth in incredible detail at: http://earth.google.com/

Keep an eye on your home while you're on holiday

You know how it is when you're away for a week or two. You have a great time but there's always that nagging little worry there could be a problem at home. When you get back you have to steel yourself before you put the key in the lock, just in case you've been burgled or a water pipe has burst.

Well, no longer. Not if you set up a webcam. Using a laptop or an internet café, you'll be able to see that your favourite antiques are still there, that the storm didn't blow the roof off, and that the neighbour is remembering to feed the cat.

For how to set up a system see *Your very own videoconferencing suite* below.

Your very own videoconferencing suite

It's quite easy to create your own videoconferencing set-up, provided you have a high-speed internet connection. Of course, it won't have the sophistication of telepresence but it could nevertheless save you quite a lot of journeys. In fact, why bother with an old-fashioned telephone at all? Get your family, friends and business contacts to do the same as you and you can all enjoy visual communication.

Here are the things that you need:

- A fairly modern computer
- A high-speed connection
- A webcam (a camera connected to your computer)
- The software to make the system work

Basic webcams are very cheap (under £20) although, as with most things, you can spend more if you want to. Normally they're attached to the computer via a cable and a USB port (some computers even have them built-in). However, if you'd like to be able to move the camera around, you'll need an extra-length video cable or, better still, some kind of cable-free set-up such as a home network.

The software will 'grab a frame' from the camera at preset intervals. In a simple system this could be every 30 seconds. But if you want people to see you moving in a natural way you'll need a rate of at least 15 frames per second (fps) and preferably 30. Naturally, the higher the rate the faster your connection needs to be.

The images need to be uploaded to a web server such as AOL Instant Messenger, Yahoo Messenger, Windows Live Messenger, Skype, Ekigg or Camfrog.

And that, basically, is all there is to it.

PART 2

Green holidays

We've arranged this part of the book in terms of geographical distance from your home. Because, broadly speaking, the less distance you travel the greener the holiday. Of course, it also matters what you do when you get to your holiday destination. But, in general, the most environmentally damaging part of any trip is the pollution produced to get there.

So we begin with holidays at home. That's right. *At home*. Can it really be a holiday? Considering how stressful holidays abroad can be (packing, putting the dog in a kennel, arranging for someone to water the garden, airport queues, delays, stomach upsets and all the rest) holidays at home can be *more* of a holiday. As long as you organise things properly. We tell you how. And, staying at home, you'll be so green that when you sunbathe on the lawn no one will know you're there.

Next we move on to holidays in Great Britain – that's to say, England, Wales and Scotland. Assuming that's where you live, you won't have to cross any water to go on holiday. And that's another important point. Because planes and ferries both produce a lot of pollution. Within Great Britain you're exceedingly unlikely to be travelling more than 500 miles (800km) one way, and probably a lot less. So don't be tempted to fly. Train, coach, a petrol-efficient car with a full load of passengers, or bicycles are the ways to get there (or even walking).

Chapter 9 takes a look at holidays on the island of Ireland. How will you get there? In fact, planes and boats have broadly the same high levels of energy use and pollution per passenger mile, so it all comes down to distance. The shortest hop is the best.

Chapter 10 describes some brilliant, green travel ideas on mainland Europe. And, finally, *Chapter 11* looks at how you can get to see the rest of the world without causing very much pollution. You might have thought long-distance travel was over. But we tell you how you can still get to see those African game parks, Asian elephants and even kangaroos.

Chapter 7

Green holidays at home

At first sight, taking a holiday at home doesn't sound a very exciting prospect. It certainly isn't going to impress the Joneses. But, then again, maybe it will. It certainly *should* impress them – with your commitment to the environment.

And when you think about it, nowhere in Britain is very far from a beach. Nowhere in Britain is very far from some attractive countryside. And just about all of the other key ingredients of a holiday are to be found everywhere in the country. If you want to go shopping, if you want to visit markets, if you want to eat out and try foreign food, if you want to see museums and art galleries, if you want to dance the night away then it's all to hand.

But it won't be a real holiday, will it?

There's no reason why not. True, a lot of people spend their 'holiday' decorating the living room or something like that. But there's no reason you have to. You can be just as lazy and carefree at home as you would have been on the Costa del Sol.

The important thing is to lay down some ground rules and stick to them. They might look something like this:

- No getting up before 9.30.
- No cooking – all meals to be eaten out.
- No housework or gardening.

- Telephone to be disconnected.
- Sunbathing every day (see below).
- Swimming every day.
- Sightseeing as the mood takes me.

Basically, think of the things you would have liked to have done on a 'normal' holiday away from home and then work at ways of doing the same things *at* home. You'll be surprised how many you can actually achieve.

WHY GO FURTHER NO.1 – Home is where the yurt is

Throughout the book we'll be repeatedly posing the question: Why go further? The fact is, we've all got used to travelling long distances simply because we can. But being green means never travelling more miles than is absolutely essential. And so we kick off with a fun idea that can literally save tonnes of pollution.

Instead of flying to Mongolia to live in a yurt, or to the Middle East to live in a Bedouin tent, or to North America to live in a tipi, set the thing up in your back garden and live the whole lifestyle. At first it sounds a bit crazy but you'll have a ball.

To start with, the accommodation. There are, believe it or not, specialist companies that can rent you everything you need. For example, Bedouin Tents of London or Arabian Tents of East Sussex [*contact details on page 98*] can create reality out of your *Son of a Sheikh* fantasies. They'll come to you and transform your lawn into a taste of the desert complete with an authentic tent and all the accessories you need including sumptuous carpets, drapes, lanterns, a hookah pipe and incense. If you feel especially romantic and in the mood for the *Kama Sutra* then there's the colourful Bombay Boudoir or, indeed, the Boho Bivouac, complete with double bed, wardrobe and sofas. Nor do you have to be constrained by reality. Let your imagination rip with the dream world of Alice in Wonderland's Tea Party or a Jousting Tournament.

Next, the food. In modern Britain there are two solutions. Either you get an appropriate takeaway or you learn the authentic cooking secrets for yourself. We favour the latter, especially as a week's course could be an integral part of your holiday. Try putting, say, 'Indian cookery courses' or 'Arab cookery classes' into your browser and see if you can find something local. Below you'll find some suggestions to get you started. Some courses are mixed, some let children take part and some are even men-only (probably because their use of the vindaloo sauce is just too dangerous for everyone else).

Nowadays it's fairly easy to get authentic ingredients in local, ethnic shops or even supermarkets. And don't forget to buy your fresh food from the organic stands at farmers' markets, whenever appropriate. You can find details of organic outlets at www.soilassociation.org Tel. 01123 145000.

As to what they eat in Mongolia, you may not find a course near you but you can read all about it at www.ub-mongolia.mn.

Finally, no Arabian tent experience would be complete without belly dancing. So another part of your holiday at home can be learning how to do it. It's fun and it's great for toning those stomach muscles.

If this all sounds a bit tame, you could use your two-week stay in the back garden to practise your new-found survival techniques. During the day you take part in a local survival course (you can find them in most parts of the country). Then you can rush home and bend one of your thankfully fast-growing *leylandii* into the frame of an emergency shelter, and start a camp fire by rubbing two sticks together.

What's green about this holiday? It's bringing the holiday destination to you rather than the other way around.

Who is it for? The whole family and maybe the neighbours too.

What's a green way of getting there? Walking out the back door.

How much pollution am I saving? If you had flown to Mongolia you would have generated about eight tonnes of carbon dioxide pollution for the return trip; for India or the Great Plains of the American West the figure would be about six tonnes.

Anything not green about it? No.

What time of year? Any time you want because yurts and tipis can be heated with a wood-burning stove or open log fire, which are carbon neutral.

What price? Weekend hire of a tipi or yurt (and even the smallest can squeeze six inside) will cost around £250 including erection, though extras like wood-burning stoves and exotic finery will add to the bill. Think £450 to £550 for a week, or even more for one of the fully-furnished jobs.

Where can I get further information/make a booking?

Tents:

→ www.bedouintents.com Tel. 07841 777268
→ www.tipihire.com Tel. 01362 680074 (Dereham, Norfolk)
→ www.hearthworks.co.uk Tel. 01749860708 (Glastonbury)
→ www.downtoearthproject.org.uk Tel. 01792 369736 (near Swansea)
→ www.bathbristoltipihire.co.uk Tel. 07591 276048
→ www.arabiantents.com Tel. 07941 370241 (East Sussex)
→ www.tipis4hire.com Tel. 07808 923341 (London and South-East)
→ www.geckoyurts.co.uk Tel. 07932 806151 (London)
→ www.worldtents.co.uk Tel. 01296 71455
→ www.thereallyinterestingtentcompany.co.uk Tel. 01803 873297

Exotic cookery classes:

→ www.renskitchen.com Tel. 07880 548227 (London)
→ www.cookforfun.shawguides.com Tel. 0207 3733651 (London)
→ www.vegsoc.org/cordonvert Tel. 0161 9252000 (Cordon Vert Cookery School, Altringham)
→ www.robrees.com Tel. 01452 770872 (Bisley Gloucestershire)
→ www.redepicurus.com or www.foodikon.com/guide/london Tel. 07900 691632 (London)
→ www.sriowen.com Tel. 0208 9467649 (Wimbledon)
→ www.vegetariancookeryschool.com Tel. 01225 789682 (Bath)
→ www.thegablesschoolofcookery.co.uk Tel. 01454 260444 (Bristol)

Other cookery information sites/day classes:

→ www.learndirect-advice.co.uk
→ www.foodofcourse.co.uk
→ www.localfoodweb.co.uk
→ www.ashburtoncookeryschool.co.uk

→ www.thanksdarling.com. Gift directory site. Thai Cookery packages.
→ www.cottagecooks.co.uk Tel. 07791 535250 Cookery classes for children (Swansea)
→ www.stirringstuff.co.uk Tel. 01491 628523 (Henley-on-Thames)
→ www.tantemarie.co.uk Tel. 01483 726957 (Woking)
→ www.leiths.com Tel. 0207 9373366 West (London)

Something organic:

→ www.uk-energy-saving.com
→ www.allorganiclinks.com
→ www.soilassociation.org.uk Tel. 01173 145000
→ www.sascotland.org Tel. 01316 662474
→ www.englishorganicwine.co.uk

Learn to belly dance:

→ www.learnbellydance.co.uk (Great Manchester)
→ www.intotheblue.co.uk Tel. 01959 578100 (nationwide)

Bushcraft and survival courses:

→ www.bushcraftexpeditions.com Tel. 01432 356700 (Hereford)
→ www.uksurvivalschool.co.uk Tel. 01432 376351 (Hereford)
→ www.survivalschool.co.uk Tel. 08712 227304 (Devon, Staffordshire, Scotland, Wales)
→ www.woodsmoke.uk.com Tel. 01900 821733 (Lake District)
→ www.survivalandsafetyschool.co.uk Tel. 01292 532372 (Scotland)
→ www.proadventure.co.uk Tel. 01978 861912 (Wales)

Don't go on holiday, move home instead

Think about it! It's not as mad as it might seem. In fact, it's not mad at all. It's totally, utterly, beautifully *sane*. Instead of living somewhere you hate and escaping a couple of times a year to somewhere you love, you simply up sticks and live where you love. And then stay put. No more holidays. At least, not of the kind that require you to travel very far.

That's what we've done. We live in Spain. To go jogging, hiking, riding or cycling we only have to open the door. To go swimming, diving, canoeing or sailing we only have to drive a few kilometres. To go skiing we only have to drive a couple of hours. We actually have no need to go on holiday at all.

The big questions you have to ask yourself are these:

- Would I need to travel *back* to the UK as frequently as I used to travel *from* the UK?
- Will friends and relatives make more foreign trips (to see me) than they used to?

If the answer is 'yes' to each question then the world isn't gaining anything. If the answer is 'no' then moving makes sense from the environmental point of view.

But, remember, you can't move to Spain (or Italy or Greece or wherever) and *still* take holidays in faraway places. You now have to take your holidays at, or close to, your new home.

It's not only beaches that are wet

As we've already pointed out, nowhere in Britain is very far from the sea. But, even so, a round trip can add up to quite a few miles for some of us, not to mention time wasted in traffic jams. And, in fact, it's not necessary because Britain has plenty of water *inland*, too. So we're suggesting you spend your holiday at home getting to know all the rivers, reservoirs, gravel pits, ponds, lakes, marshes and bogs near you.

Let's start with water from the point of view of wildlife. Down on the coast you may have been hoping to see gulls and terns and ducks and geese but inland you'll see them, too. They're pretty much at home in both places. But inland you may also see herons, harriers, bitterns, kingfishers, voles, otters ... plenty of things. Even in deepest Warwickshire, which is about as far from the coast as you can get, there are ex-waste-tip pondlands. Put 'wildlife reserves' into your search engine together with the name of your county and see what it comes up with. Or drop in at your nearest tourist office – it's not just for foreigners. You'll probably be surprised.

Here's one place you probably never would have expected to find wading birds:

- **The London Wetland Centre**. It's not, as you might have thought, miles out on the Thames marshes. It's actually in Barnes, which is pretty central. Created from a disused Victorian reservoir, the amazing 100-acre site includes a purpose-built collection of different types of wetland habitats, such as reedbeds, wet woodland, open water and grazing marsh, all of which can be viewed from boardwalks and hides. Here you'll be able to see the bittern, which used to be a rarity. What's more, the Wildfowl and Wetland Trust which runs the centre (and many others all over the country) has introduced high technology so you can see, for example, close-ups of nests on monitors.

And here are a few other bogs and wetlands to check out:

- Moseley Bog, Birmingham
- Crymlyn Bog, Swansea
- Muir of Dinnet, Aberdeen
- Thursley and Ockley Bogs, Guildford
- Malham Tarn, Skipton
- Irthington Mires, Hexham
- Garry Bog, Coleraine, County Antrim
- The Outer Ards, Bangor, County Down

Altogether, about 150 soggy places up and down the country are recognised by the RAMSAR Convention on Wetlands of International Importance (so called because it was signed in Ramsar, Iran in 1971). You can find the nearest one to you by looking at www.jncc.gov.uk/pdf/ris. The Convention is supposed to protect the wetlands but in the face of development and government indifference it often fails. So it's important to visit wetlands – probably the least appreciated landscape – to show you care.

The other thing you'd probably like to do in inland waters is swim. The first problem is the fear that inland waters can be cold. Not a bit of it. Why, the famous swimmer Lewis Gordon Pugh has pronounced them like a bath. Mind you, he swam the Arctic at zero degrees and the Antarctic at minus 1.7 degrees. If you're not quite so tough you might like to consider one of the wetsuits specially made for swimmers. They start at under £100 and, with care, will last for several years. What's more, because of their buoyancy, they'll improve your speed and probably your style as well.

The other problem is that in England and Wales only about two per cent of the length of rivers is accessible to the public. That's in complete contrast to Scotland where, since 2003, right of access has been guaranteed by law. You can find suggestions for legal sites south of the border at www.river-swimming.co.uk, the website of the River and Lake Swimming Association. If you think there should be more of them then join (it's free) and add your voice to their campaign.

What's green about this holiday? It's local and it's giving value to landscapes that are often the least appreciated.

Who is it for? The whole family.

What's a green way of getting there? Bus, bike or on foot if it's close enough.

How much pollution am I saving? That depends on where you'd normally go on holiday but by not flying to, say, the south of France you'd be saving one tonne of carbon dioxide.

Anything not green about it? No.

What time of year? For waterfowl winter; for swimming summer.

What price? Usually free, although there may be a charge for guided tours. But, in any case, always make a donation.

Where can I get further information?

➜ www.birmingham.gov.uk Tel. 0121 464 8728

➜ www.ramsar.org

➜ www.wwt.org.uk Tel. 01435 891900

➜ www.wildlifeextra.com (online magazine)

➜ www.rspb.org/reserves/area

➜ www.enjoyengland.com

➜ www.naturalengland.org.uk

➜ www.english-nature.org.uk

Become a tourist in your own town

When you live somewhere you often know less about it than a tourist. So here we're suggesting that for your deeply green holiday this year you become a tourist in your own town. Dig out the camera, brush off the day pack, put on the shorts and buy a map. You'll find it's far more fascinating than you ever imagined.

First stop is the local tourist office to get some tips and pick up some information. Here's a taster of the kinds of things you might discover:

- **Wiltshire**. You'll already know all about Stonehenge, of course. But do you also know about Old Sarum, a fascinating archaeological site just two miles north of Salisbury and a human settlement from around 300 BCE? It was here, according to most historians, that the first Salisbury Cathedral was built. Work was finished around 1092 but within the week it had burned to the ground. You might then like to move on to the present Salisbury Cathedral which has the tallest spire in England. But did you also know that the cathedral houses the world's oldest working clock? Oh, and Salisbury was the birthplace of *Some Mothers Do Have 'Em* actor Michael Crawford.
- **Southport**. Bet you haven't been inside the Lawnmower Museum in Shakespeare Street. It may sound a bit of a joke and, indeed, the inventor, Edwin Beard Budding faced so much ridicule back in 1830 that, according to legend, he had to carry out tests on his machine after dark. But just think where we'd be without it. Budding worked in the textile mills where the machines used to trim the nap of material gave him the idea.
- **Norfolk Broads**. There's much more to the Norfolk Broads than the Broads. Have you, for example, visited the Gressenhall Workhouse in Dereham? It only stopped being a place of refuge in 1948, a salutory reminder of what life was like without the welfare state. The building houses a museum and is set in the grounds of a rare breeds farm which is also open to the public.
- **Buckinghamshire**. You and just about everybody else in the country knows about the Bletchley Park Museum near Milton Keynes with its

famous code-breaking enigma machine. But much less-known is the American Garden outside the building. Established with the help of the Royal Horticultural Society, it commemorates the working relationship during and after the war between the UK and US military. The walk starts with the Giant Sequoia Tree (emblem of California), moves on past a giant cactus (Arizona), a Lilac (New Hampshire) and so on.

- **Edinburgh**. Have you yet tried *The Underground*? No, it's not a tube service but a new eating and entertainment complex deep under the streets of the city around the South Bridge. Formerly the vaults which had lain mostly abandoned for decades, it now has nightly ceilidhs, traditional Scottish entertainment and, even, ghost walks. The 'callers' (the people who tell you what to do to the music) come from all over the country and even if you know the Edinburgh form of 'strip-the-willow' you'll find the Orkney and Shetland versions are very different.
- **Lincolnshire.** The county is renowned, amongst other things, for Lincoln Cathedral, Fenlands and, of course, Margaret Thatcher who was born in Grantham. You know all that. But did you know about the Pinchbeck Engine and Land Drainage Museum near Spalding? It might sound a bit dull but actually as the last (restored) steam pumping station in the South Fenlands it's of international importance to the history of land drainage. Later, pop into the Gordon Boswell Romany Museum in Spalding, one of Europe's largest collections of all things Romany, from vardos (the typical caravan) to carts and all types of harness. There are also horse-drawn excursions (weekends only). In the north of the county in a village called Alkborough there's a 43-foot (13 metre) circular turf maze known as Julian's Bower. Its exact origin is unknown. It could be 16th century, it could even be Roman. Who says they never did anything for us! Finally, Grimsthorpe Castle close to the town of Bourne, was the filming location for *The Fortunes and Misfortunes of Moll Flanders*. Maybe one day the location scouts will be looking for a place to film the life story of that other famous lady.
- **Shropshire**. Ironbridge Gorge, if you didn't know, is a World Heritage

Site and there are at least ten museums in the area. But the quirkiest fact is that the gorge valley is also the birthplace of the famous Aga cooker and one of the Aga factories is still operational in the village of Coalbrookdale. You can't actually go into the factory, but one or other of the museums in the area often has exhibits and special displays commemorating the beginning of this great British institution. And there are cookery demonstrations, too.

- **Gloucestershire**. Bet you've never heard of Hetty Pegler's Tump – no, it's not in the least bit rude. It's actually the local name for Uley Tumulus near Dursley in the Cotwolds, and comes from the Anglo-Saxon for mound. An impressive megalithic monument, it's very little visited but much respected by those who know about these sorts of things. At Stroud, known as the 'Covent Garden of the Cotswolds', there's the Painswick Rococo Garden, an unusual blend of foliage and rococo architecture which was saved by a painting. Established in the 1740s the flamboyant gardens had fallen into disrepair by the late 19th century. In the 1980s it was decided to restore the gardens to their former glory. But how could anyone know how they looked two centuries earlier? The answer was in a series of contemporary oils by Thomas Robins. And so the work was done.
- **Herefordshire**. There's another tump here, the Wormelow Tump. It's also the site of the Violette Szabo museum, set up in the home of the family of this World War II heroine as depicted in the book and film *Carve Her Name With Pride*.

What's green about this holiday? It's local.

Who is it for? The whole family.

What's a green way of getting there? Bus, bike or on foot if it's close enough.

How much pollution am I saving? That depends on where you'd normally go on holiday but by not flying to, say, Italy you'd be saving around 1.5 tonnes of carbon dioxide.

Anything not green about it? No.

What time of year? Any time.

What price? You'll only have to pay for fares and entrance charges plus a few meals out and whatever other entertainments you fancy; around £100 each should do it.

Where can I get further information? Just drop in at your nearest tourist office or begin your researches at the following websites:

- ➜ www.enjoyengland.com Tel. 0208 8469000
- ➜ www.visitwales.com
- ➜ www.visitscotland.com
- ➜ www.discovernorthernireland.com
- ➜ www.visitbritain.org
- ➜ www.visitlincolnshire.com
- ➜ www.hiddenengland.org
- ➜ www.grimsthorpe.co.uk Tel. 01778 591205
- ➜ www.sholland.gov.uk Tel. 01775 725468 (Pinchbeck)
- ➜ www.boswell-romany-museum.com Tel. 01775 710599
- ➜ www.bletchleypark.org.uk Tel. 01908 640404
- ➜ www.undergroundedinburgh.co.uk Tel. 01312255440
- ➜ www.aboutbritain.com.towns/alkborough.asp
- ➜ www.ironbridge.org.uk Tel. 01952 433970
- ➜ www.visitthecotswolds.org.uk Tel. 01453 760960
- ➜ www.rococogarden.org.uk Tel. 01452 813204
- ➜ www.violette-szabo-museum.co.uk Tel. 01981 540477

WHY GO FURTHER NO. 2 – Autumn colours

New England and south-eastern Canada have become world famous for the autumn colours of their maples. And, indeed, they're spectacular. But no more so, in fact, than many of the autumn displays we have on our doorsteps. It's all a question of marketing.

The Westonbirt Arboretum near Tetbury in Gloucestershire, for example, is just one of the so-called five 'Great Ws'. Managed by the Forestry Commission, it's an *aceretum.* That's to say it's planted with acers, the technical name for maples. In fact, it's home to the National Collection of Japanese Maples comprising about 18,000 specimens. So, come the autumn, it's pretty impressive. The Japanese obviously know a thing or two about autumn's colours. They actually have a name for the business of going out specially to look at the displays. So ask the kids if they fancy a bit of *momijigari*. It beats *tamagotchi* any day.

The other members of the 'Great Ws' are:

- Wakehurst Place Garden
- Windsor Great Park
- Winkworth Arboretum
- Wisley Garden

If you go to the Forestry Commission website [*see below*] it will actually tell you when the displays are reaching their peak and give details of how to get there. The Woodland Trust's website does the same. Here are some of its best autumn colours:

- Swineshead and Spenoak Wood, Bedfordshire
- Penn Wood, Amersham, Buckinghamshire
- Milltown and Lantyan Wood, near Lostwithiel, Cornwall
- Milton Rigg Woods, Brampton, Cumbria
- Shepton Woods, Bovey Tracey, Devon
- Pontburn Woods, Rowlands Gill, Durham
- Hainault Forest, Romford, Essex ('great colours and four *great* pubs nearby')
- Lineover Woods, Dowdeswell, Gloucestershire
- Valley Park Woods, Chandler's Ford, Hampshire ('local pub')
- Hucking Estate, Hollingbourne, Kent
- Bilton Beck and Rudding Bottoms, Scotten, North Yorkshire
- Stoke Wood, Stoke Lyne, Oxfordshire
- Marden Park, Woldingham, Surrey
- Galgorm Wood, Co. Antrim
- Killaloo Wood, Co. Londonderry
- Abriacham Wood, Scottish Highlands (view over Loch Ness)

- Crinam Woods, Argyll and Bute, Scotland (views over Mull and Jura)
- Coed Ystrad, Johnstown, Carmarthenshire
- Green Castle Woods, Llangain, Carmarthenshire
- Coed Aber Artro, Gwynedd

If you live in or near London you also have the privilege of being able to shuffle through the leaves in the Royal Parks:

- Bushy Park
- Green Park
- Greenwich Park
- Hyde Park
- Kensington Gardens
- Regent's Park
- Richmond Park
- St James's Park

And don't forget the Royal Horticultural Society Garden, Kew, whose 300 acres are a UNESCO World Heritage Site. London's parks often come with music and entertainment as well and, in the case of Kew, an ice rink.

Finally, it's not just a question of trees. Many gardens are also designed with autumn colours in mind, as you'll see on the Great British Gardens website [*below*].

The national what?

If, like us, you have rather lost track of what's going on with the creation of the UK National Forest here's a bit of an update. The project, which came to life as a vague idea from the Countryside Commission in 1987, was officially launched in 1991 and involves transforming 200 square miles of England's central counties (Leicestershire, Staffordshire and Derbyshire – at present amongst the least densely wooded areas of the country). Essentially, the gaps between the existing ancient forests from Charnwood in the east to Needwood in the west are to be filled in, encircling the towns of Burton upon Trent, Coalville, Swadlingcote and Ashby de la Zouch.

So far 17 per cent of the area is woodland (90 per cent of which is open to the public). This compares to six per cent in 1991 and with a target of a third of the total area when the forest is 'finished'. So far, seven million trees have been planted, many on obsolete coal and mineral mining areas and some farmland has been converted. As the National Forest Company points out, you too can take part in the project by paying £25 to have your own (or gift) tree planted and receive a certificate to prove it. Not a bad idea we think for Christmas, birthdays and other special occasions. And you don't even need to sharpen your spade.

What's green about this holiday? It's local and it's supporting trees.

Who is it for? Anybody who loves getting out in the autumn, shuffling through the leaves, maybe even playing conkers and then getting back for a nice, cup of hot chocolate.

What's a green way of getting there? Bus, bicycle or on foot, depending on the distance.

How much pollution am I saving? Compared with flying to Canada to see the maples you'll be saving around 4.5 tonnes of carbon dioxide.

Anything not green about it? No.

What time of year? Autumn.

What price? Some of the more elaborate gardens, such as Kew, charge an entrance fee, but most city and town parks do not. In any case, you should always make a donation for the upkeep and conservation of something very special.

Where can I get further information?

➜ www.royalparks.gov.uk Tel. 0207 298 2000

➜ www.greatbritishgardens.co.uk Tel. 01666 825390

➜ www.visitnationalforest.co.uk Tel. 01283 551211

➜ www.woodland-trust.org.uk Tel. 0800 0269650

- → www.forestry.gov.uk Tel. 0845 3673787
- → www.kew.org Tel. 0208 3325000
- → Batsford Arbetorum, Moreton-in-Marsh, Gloucestershire, www.batsarb.co.uk Tel. 01386 701441
- → Dawyk Botanic Garden, Peebleshire, Scotland, www.rbge.org.uk Tel. 01721 760254
- → Cambridge Botanic Garden, www.botanic.cam.ac.uk Tel. 01223 336265
- → Castlewellan National Arbetorum, Northern Ireland, www.forestserviceni.gov.uk Tel. 02843 778664.
- → Inverewe Gardens, Poolewe, Ross-shire Tel. 01445 781200
- → Sir Harold Hillier Gardens, Romsey, Hampshire www.hillier.hants.gov.uk Tel. 01794 368787
- → Localsh Woodland Garden, Balmacara, Ross-shire Tel. 01599 566325
- → Sezincote, Moreton-in-Marsh, Gloucestershire www.sezincote.co.uk.
- → Sheffield Park, Uckfield, Sussex Tel. 01825 790231
- → RHS Wisley, Woking, Surrey www.rhs.org.uk Tel. 01483 224234
- → Wakehurst Place, Haywards Heath, West Sussex www.kew.org.uk Tel. 01444 894066

And finally, for alerts about the 'wrong kinds of leaves' falling on the railway lines click on: www.nationalrail.co.uk

Your green holiday home

This is probably the greenest thing you can do for a holiday. Not just stay at home but, at the same time, transform it into the most ecological abode that's possible. It makes good sense from every point of view. Instead of *spending* money on a holiday you *invest* it in things that make you more comfortable, save you money in the future, add to the value of your home and help save the planet.

Number one improvement has got to be *energy saving*. Environmentalists call this the 'fifth fuel'. At its simplest it just means turning off unnecessary lights and anything on standby, not running the water when you clean your teeth, showering instead of bathing, and so on. Those things are free but with your holiday money you can do a lot more.

Here are some facts:

- External walls can lose 35 per cent of heat without efficient insulation
- Roofs and ground floors can lose 15 per cent of heat
- Energy-saving light bulbs should pay for themselves within one year
- Sheep's wool insulation only takes 14 per cent of the energy necessary to manufacture glass-fibre insulation.

The first step is to improve your knowledge and your skills. So the initial week of your holiday will be an 'E-DIY' course. One of these may be near to you:

- **The Low Impact Living Initiative** (LiLi) at the Redfield Community, Winslow, Buckinghamshire. Here you can learn to make natural paints and lime-washes. Or you can buy their Aglaia range of paints which uses no petrochemicals and whose waste materials are totally compostable. You'll get plenty of natural and home-grown stuff inside you, too, as Redfield will feed you and, if you need it, give you a bed for the night. There's a car-sharing scheme to help you get there.
- **Green Dragon Energy** is a German-based company running courses all over the country on alternative energy and how to apply it to your life. They'll take you through the options – solar, wind, thermal – and also give you hands-on experience.
- **The Earthworks Trust** runs the Sustainability Centre in Petersfield, Hampshire, where there are courses on just about every aspect of alternative living.
- **Plant Dyed Wool**, Hereford, will teach you all about home-brewed natural dyes so that, instead of buying new furniture and lining Mr Ikea's pockets, you can rejuvenate your existing furniture and furnishings and introduce a new colour-scheme.

- **Mourne Textiles** in the Mourne region of Northern Ireland, and **Trigonos**, Snowdonia, will both teach you how to weave your own rugs from organic wool.
- **Lilac Barn**, Somerset, will teach you how to do your own upholstery, while **Paragon Courses**, Thaxted, Essex will show you everything you need to know about furniture restoration.
- **The Organic Centre** in Rossinver, County Leitrim, and the **Cutting Garden** in Robertsbridge, East Sussex, have organic gardening courses for beginners where you can learn all about ground planning, ground preparation, rotation and soil fertility management.
- If the garden fence blew down in the last gales and you'd like to replace it with something more permanent, take the **Derby College**, Ilkeston, dry stone walling course.

With your courses behind you, you're now ready to start work. Improving the insulation of your home is probably the single most important thing you can do. Go, wherever possible, for a material derived from organic sources. The best eco-suppliers will be able to offer such things as *Thermafleece,* (thermal insulation from British hill sheeps' wool), *Warmcell* (blown fibre loft insulation), *Isonat* (natural insulation made from hemp fibre and recycled cotton) and *Rockwool,* which is mainly made from volcanic rock and which – its manufacturers claim – saves '100-fold' more greenhouse gases than are used in its production.

When it comes to brightening things up afterwards you should avoid regular paints (as you'll have learned at Redfield). They all contain, amongst many other things, solvents and petro-chemical based additives. Eco-varieties try to ensure that their ingredients are as natural and harmless as possible. (But bear in mind that 'natural' can include toxic substances like lead, arsenic and titanium dioxide – mined and processed at considerable environmental cost.) The best advice is to buy from a recognised green specialist. Eco-paints tend to be more expensive than their industrial counterparts and job preparation such as the cleaning of woodwork has to be more thorough. Interestingly, though, because of their natural ingredients like linseed oil, the application of eco-paint should be beneficial to the surfaces.

What's green about this holiday? You're not only staying at home, you're also reducing your home's impact on the environment. So it's a double-whammy.

Who is it for? Hopefully everybody in the family will join in.

What's a green way of getting there? You are there. For the courses, find the most local you can.

How much pollution am I saving? By not flying to, say, Greece you'll be saving about 2.5 tonnes of carbon dioxide. By improving your home you could easily save another tonne a year.

Anything not green about it? No.

What time of year? Any time

What price? The courses mentioned tend to range from about £60 for the one-day organic gardening lessons in Rossinver up to £200 for a two-day course, usually including some food but not accommodation. As to the cost of the home improvements, it's up to you but to give you an idea, you could probably insulate your loft for about £300.

Where can I get further information?

→ www.cat.org.uk Tel. 01654 705950 Powys, Wales
→ www.aecb.net (Association of Environment Conscious Building). Links to courses, seminars and forums. Tel. 0845 4569773, Llandysul, Wales.
→ www.lowcarbonkid.blogspot.com (informal information)
→ www.businessgreen.com
→ www.carbontrust.co.uk Tel. 0800 0852005
→ www.newconsumer.com
→ www.allthingsgreen.net
→ www.houseplanner.co.uk
→ www.buildstore.co.uk Online building and DIY magazine with many alternative ideas. Links to the Big Green Home Show held at the National Self- Build and Renovation Centre in Swindon. Tel. 01506 409616
→ www.homebuildingshow.co.uk Birmingham, April; Glasgow, May; Newbury, June; London, September; Somerset, November; and Harrogate, November.

Green insulation products:

→ www.insulation-actis.com
→ www.celotex.co.uk
→ www.knaufinsulation.co.uk
→ www.sigplc.co.uk
→ www.ybsinsulation.com

Suppliers of 'green' mortar and other eco-products:

→ www.hempshop.co.uk
→ www.greenshop.co.uk

Eco-paints suppliers:

→ www.ecopaints.co.uk Tel. 0845 3457725
→ www.nutshellpaints.com (Devon) Tel. 01364 73801
→ www.lime.org.uk (Calch-Ty-Mawr, Brecon) Tel. 01874 658249
→ www.livos.com Tel. 01795 530130
→ (Osterman&Scheiwe, Buckinghamshire) No web. Tel. 01296 481220
→ (Potmolen Paints, Wiltshire) No web. Tel. 01985 213960
→ www.strip-paint.com (Eco Solutions, Somerset) Tel. 01934 844484
→ www.ecospaints.com (Lakeland Paints, Cumbria) Tel. 01539 732866

Natural building products:

→ www.oldhousestore.co.uk
→ www.greenbuildingstore.co.uk
→ www.constructionresources.com (London) Tel. 0207 4502211

Photovoltaics:

→ www.nu-loc.com/uk

Sustainable homes:

→ www.eco-hometec.co.uk
→ www.sustainablebuildingsolutions.co.uk
→ www.ecomerchant.co.uk

Eco-build courses:

→ www.lowimpact.org Tel. 01296 714184
→ www.soilassociation.org/masterclasses
→ www.cat.org.uk Tel. 01654 705950 Powys, Wales
→ www.greendragonenergy.co.uk Tel. 00 49 3048624998 (Berlin-based renewable energy company running courses all over the UK)

Interior and exterior courses:

→ www.holidaycourses.co.uk General directory
→ www.plantdyedwool.co.uk Tel. 01981 590370
→ www.adcrafts.info/mournetextiles.htm Tel. 02841 738373
→ www.paragoncourses.co.uk Tel. 01371 832032
→ www.lilacbarn.co.uk Tel. 01823 690134
→ www.theorganiccentre.ie Tel. 00 353 719854338
→ www.thecuttinggarden.com Tel. 0845 0504849
→ www.derby-college.ac.uk Tel. 01332 520200

Chapter 8

Green holidays in Britain

The words 'holiday' and 'abroad' nowadays tend to go together like fish and chips. Especially when it comes to things like sunbathing, heat, watersports, vineyards, food for foodies, camping and skiing. But, in fact, most of the things we go abroad for are just as good in Britain. You'd be amazed. Why go further?

WHY GO FURTHER NO. 3 – The suntan

Coming back with a suntan has always been one of the aims of the foreign holiday. But, in fact, you can get a perfectly good real suntan in Britain. And it'll be a much *safer* tan than one from the Mediterranean. The lobster-skinned appearance of north Europeans is a common and slightly comical sight around Mediterranean beaches in summer. But when you consider that red skin is *literally* burnt it's not as funny as all that.

Of course, a tan is the body's way of trying to protect your skin. So, logically, you shouldn't try to tan faster than your body's ability to adjust. That means starting off with just a few minutes a day. But, of course, you're going to be in a hurry. You don't want to be all white when you strip down to your bathing gear. So what can you do?

Well, you could take a course of sunbed treatments, beginning a few weeks before the holiday. But take note that, according to the World Health Organisation (WHO), sunbeds are *not* safer than the real thing. A lot of exposure increases your chance of developing malignant melanoma, a form of skin cancer. You're also at risk of damaging your eyes and ageing your skin prematurely. And you may not know until it's too late because symptoms can take up to 20 years to appear.

According to the WHO you should never use sunbeds:

- More than twice a week nor for more than 30 weeks a year
- If you're a teenager or child (your skin is much more sensitive)
- If you're very fair-skinned, freckled or have moles
- If you've had skin cancer or have a family history of skin cancer
- If you're taking medication that makes your skin more sensitive to ultraviolet rays

Expect to pay around £10 for initial, short sunbed sessions and £20 for the longer ones that come later.

For further advice:
→ www.sunbedassociation.org.uk
→ Tel. 01494 785941

Another idea is the tan in a bottle, which has come a long way since the smelly beginnings with built-in streaks. Home tans now give a pretty good result but professional jobs are better. You can have it massaged in, sprayed by hand or – like a car in a paint shop – sprayed by robot in a walk-through cubicle. Expect to pay around £7-£10 for a home fake tan cream and £25-£50 for a salon job. On top, there are always those little extras such as pre-tan moisturisers, post-tan moisturisers, tan enhancers and tanning glitter powders (and they're not just for the girls).

Bear in mind that most fake tans *don't* give any sun protection, so you're going to need to apply that on top every time you're on the beach.

Further information:
→ www.thetanningshop.co.uk Tel. 01325 636900
→ www.gorgeousshop.co.uk
→ www.sunless.com
→ www.sunlabs-uk.com

So you've got some sort of a pre-tan and you've got your sun protection.

Now what? Obviously you're champing at the bit to get to the sunniest beach you can find. But not so fast. When it comes to sunbathing, less is very definitely more. The World Health Organisation has developed a UV Index and it's now shown on weather forecasts as a number inside a triangle. You can see the full details at http://info.cancerresearch.uk.org but in broad terms it can be said that for *average* skin:

- UV levels 1 to 4 – safe
- UV levels 5 to 6 – keep out of the sun between 11am and 3pm and apply factor 15
- UV levels 7 to 10 – dangerous

On a clear summer's day in Britain the UV level is normally no higher than seven around noon. But in places like Spain it can actually reach 10, the highest. Which means you've travelled a long way to sit under an umbrella or indoors most of the day.

And, in fact, the record for sunshine hours along England's south coast, the east coast up to the Wash, the Severn Estuary and Pembrokeshire is an impressive 1,541 to 1,885 hours a year, only a quarter less than for Barcelona.

It comes down to this. England and Wales have holiday destinations with plenty of sunshine in which it's safe to sunbathe sensibly. The Mediterranean has more sunshine but, in summer, it's frankly just too dangerous.

What's green about this holiday? You'll be staying in the UK.

Who is it for? British UV levels are much safer for children and those with fair skins.

What's a green way of getting there? Coach, train or energy-efficient car with all seats taken; if the car isn't full, use Freewheelers to offer the spare places (www.freewheelers.com).

How much pollution am I saving? Compared with flying to, say, a Spanish *costa,* you'll be saving about one tonne of carbon dioxide.

Anything not green about it? Lying on a beach is pretty harmless.

What time of year? The traditional August break is pretty good – by then the UV levels are generally down from the peak.

What price? The range is huge.

Where can I get further information? The Met Office website will tell you the UV level: www.metoffice.gov.uk

WHY GO FURTHER NO. 4 – Surfing

You're lying flat on your board, eyes and ears tuned to the critical moment when you must start to arm-paddle swiftly enough to synchronise with the upcoming wave. Now you can feel it. Right, this is the moment to kneel fluidly and then crouch to ride that wave. Yeeeees. But this isn't Portugal, or Australia, but Newquay. So how do they all compare?

The first thing to point out is that if you're a beginner you're hardly going to be looking for those famous 'barrels' of water that professionals are always photographed in. That would be like climbing Mount Everest for your first ski lesson. What you want is high-quality tuition – in English. And at British schools that's what you'll get. Your first time, your instructor will be right in there beside you telling you when the moment has arrived and, most probably, lifting the board (usually foam for beginners) off your head so you can try again.

But what about once you've progressed? In fact, many of this country's best surfing areas are very very cool indeed – even in international terms. The Annual Rip Curl Boardmasters competition at Newquay is a seriously world-famous event. For one whole week in August, the area resembles Bondi Beach as the experts vie for $100,000 prizes. The surf is that good. Non-surfing girlfriends meanwhile enjoy competitions such as Bikini Babe of the Year. But don't for one minute think that's all the gals get up to. More and more women are taking to the waves and many of the surf schools offer women-only sessions (probably because the men can't stand the competition).

According to the latest estimates, some 300,000 Brits already know how to surf, are hooked and are very happy with British conditions. And why not? That's the Atlantic out there. It's serious. July/August is the holiday peak and the most popular with beginners. But come the autumn the water is warmer and the first of the winter storms way out across the Pond start to build swells with a fetch of hundreds of miles.

About half the industry's turnover is generated in the south-west but the best surf is often to be found in Scotland, especially Dunbar in Perthshire and in the Hebrides. It may not be sunny but once you're good that's the last thing you'll be looking for. You'll want a bit of froth on your waves and you'll need a good wetsuit, of course. Conditions thrust Scotland pretty close to the Top Ten in the world and there are no crowds, either.

In between Cape Wrath and Land's End other good areas include Pembrokeshire and, for the bright, young adrenalin-junkies, the Severn Bore.

What's green about this holiday? You won't need to travel very far and you'll be riding the ultimate, ecological wave-powered boat.

Who is it for? It helps to be young – but it's not mandatory.

What's a green way of getting there? The Big Friday Bus [*see below*] is the ecological solution from London. Otherwise, any train or coach or fuel-efficient car with all seats taken.

How much pollution am I saving? Compared with flying to Australia you'll be saving about 10 tonnes of carbon dioxide, roughly equivalent to an entire year's normal carbon footprint.

Anything not green about it? The board had to be made, so if you're feeling really green you could dispense with it and body surf.

What time of year? Spring if you're desperate, summer if you're there for the party, autumn for the combination of warm sea and big waves and, yes, winter if you're mad.

What price? From £200-£300 for a week's surfing lessons with your transport, food and accommodation on top.

Where can I get further information?

→ www.britsurf.co.uk Tel. 01637 876474

→ www.surfing-waves.com

→ www.surfingholidays.net
→ www.kneeboardsurfing.co.uk
→ www.kneelo.org
→ www.body-surfing.co.uk
→ www.a1surf.com (forecast)
→ www.magicseaweed.com (forecast)
→ www.bbc.co.uk/weather/coast/surf/ (forecast)
→ www.surfsizenow.com (forecast)
→ www.ripcurlboardmasters.com

How do I make a booking?

→ www.bigfriday.com Tel. 01637 872512
→ www.surfingcroydebay.co.uk Tel. 01271 891200 (Devon)
→ www.outdooradventure.co.uk Tel. 01288 362900 (Cornwall)
→ www.hebrideansurf.co.uk Tel. 01851 840337 (Isle of Lewis)
→ www.surfschoolscotland.co.uk Tel. 07793 063849 (Inner Hebrides)
→ www.wsfsurfschool.co.uk Tel. 01792 386426 (Wales)

WHY GO FURTHER NO. 5 – White water rafting ... and bugging

When somebody says "hold-on" you know you're in for a thrill – and there's no getting off, well, not elegantly anyway. Going down the River Tummel in Scotland is something like being in the drum of a washing machine on a cold cycle. There are plenty of 'fun technical rapids' with scary names like Zig Zag, Bone Yard and Shark Tooth, and the whole thing ends with an 18-foot drop. Well, they did tell you to hang on.

So there's no need to travel to the Grand Canyon or Zell am See. Britain can compete with many of the more exotic-sounding rivers of the world. And another plus is that when somebody up front shouts "paddle to the left like your life depends on it" you don't need a translation.

To white water rafting enthusiasts, rivers come in six strengths. One is as placid as a stirred cup of tea while six is the froth-maker in a coffee machine. Britain has all types – think two to five for the River Tummel. But if you think you'd prefer to start with a spot of Level One then it's the River Tay for you.

Get bugging

Ever wanted to call yourself – or somebody else – a real bugger? Well, now you can, without a moment's blushing, because there's a whole new whitewater sport called *bugging.* Bugs are a cross between a raft and an armchair. Inflatable chambers ensure you stay on top of the water while extra thick pads allow you and your craft to crash through rapids without getting hurt. Well, that's the theory. Operators claim that with a little practice you'll be able to run the rapids solo, catch eddies, surf standing waves and even pirouette. Most bugging trips come in half-day packages for about £50 all in, but they are restricted to the over-16s. See www.naelimits.co.uk Tel. 08450 178177.

What's green about this holiday? Apart from your face? It's in Britain, rather than the USA or New Zealand. And it doesn't use a motor.

Who is it for? You need to be a bit daring. Most operators will insist on adult supervision of anybody under 18, and don't allow under-16s on board for the whiter stuff. But on the quieter stretches it's usually OK for the over-12s.

What's a green way of getting there? Coach, train or energy-efficient car with all seats taken; if the car isn't full, use Freewheelers to offer the spare places (www.freewheelers.com).

How much pollution am I saving? By not flying to the USA for the Grand Canyon you'll be saving around six tonnes of carbon dioxide; by not flying to New Zealand you'll be saving about 11 tonnes.

Anything not green about it? It's not the most tranquil use of a river.

What time of year? Depends on the river but most locations will have enough water during the summer and more than enough in spring and autumn.

What price? Half-day trips are around £50. A week-long multi-location package will be around £500, bed and breakfast.

Where can I get further information? Always look for the AALA badge (Adventure Activity Licensing Authority) when booking white knuckle stuff. They cover England, Scotland and Wales. See www.aala.org Tel. 01292 0755715.

Also:

→ www.visitscotland.com Tel. 0845 2255121
→ www.visitwales.co.uk Tel. 08708 300306
→ www.wales1000things.com
→ www.discoverdevon.com Tel. 0870 608553
→ www.cornwalltouristboard.co.uk Tel. 01872 322900
→ www.bluedome.co.uk
→ www.sportandadventure.co.uk
→ www.exelement.co.uk

How do I make a booking?

→ www.aceadventure.co.uk Tel. 01479 810510
→ www.activitywales.com Tel. 01437 765777
→ www.verticaldescents.com Tel. 01855 821593
→ www.naelimits.co.uk Tel. 08450 178177
→ www.adrenalunatics.com Tel. 01654 767655
→ www.riverdeepmountainhigh.co.uk Tel. 015395 31116
→ www.riverdart.co.uk Tel. 01887 829706
→ www.watercooled-adventures.co.uk Tel. 07901 868191
→ www.wye-pursuits.co.uk Tel. 01600 891199

WHY GO FURTHER NO. 6 – Canyoning

Before we leave the subject of water we must mention canyoning, a sport strongly associated with the limestone massifs of the Pyrenees, the Cevennes and parts of the USA. If you ever saw a film called *The Beach* you'll have an idea. Remember that scene where Leonardo DiCaprio slides down a waterfall into a cauldron of blue water? Well, that's canyoning. In other words, following a river down through a steep gorge, sometimes wading, sometimes swimming, sometimes sliding and sometimes jumping.

But to enjoy those thrills you don't actually have to go any further than Fort William in Scotland or Snowdonia in Wales. And there's even a bit going on in the Lake District.

A few words of advice. However tough you are, you must go with a guide who knows the river. Otherwise you just can't tell where it's safe to jump in and where there are hidden rocks. Some people jump straight as arrows and go in deep. But we were taught to jump with legs like a frog and the nose firmly pinched between thumb and forefinger. That way you don't go in deep and the water doesn't get firehosed up your nostrils. Of course, you have to wear a wetsuit.

What's green about this holiday? It's in Britain, it doesn't use a motor and it's about the only way of appreciating what would otherwise be a hidden landscape.

Who is it for? There are canyons for all the family and others for hard men (and women) only.

What's a green way of getting there? Coach, train or energy-efficient car with all seats taken; if the car isn't full, use Freewheelers to offer the spare places (www.freewheelers.com). Your guide should arrange local transport by fuel-efficient minibus.

How much pollution am I saving? By not flying to the Pyrenees you'll be saving around a tonne of carbon dioxide.

Anything not green about it? No – as long as you take care not to damage vegetation.

What time of year? That depends entirely on the local river system. Spring and autumn can have too much water and summer too little, so keep in close touch with your guide beforehand. Above all, never, ever, be tempted to canyon when heavy rain is forecast.

What price? A full-day will be in the region of £40-£60 usually including the cost of wetsuit and helmet hire and local transport. A seven-day package will set you back about £500 bed and breakfast.

Where can I get further information? Always look for the AALA badge (Adventure Activity Licensing Authority) when booking white knuckle stuff. They cover England, Scotland and Wales. See www.aala.org Tel. 01292 0755715.

Also:

→ www.visitscotland.com Tel. 0845 2255121

→ www.visitwales.co.uk Tel. 08708 300306

→ www.wales1000things.com

→ www.discoverdevon.com Tel. 0870 608553
→ www.cornwalltouristboard.co.uk Tel. 01872 322900
→ www.bluedome.co.uk
→ www.sportandadventure.co.uk
→ www.exelement.co.uk

How do I make a booking?
→ www.verticaldescents.com Tel. 01855 821593
→ www.naelimits.co.uk Tel. 08450 178177
→ www.adrenalunatics.com Tel. 01654 767655
→ www.riverdeepmountainhigh.co.uk Tel. 015395 31116
→ www.riverdart.co.uk Tel. 01887 829706
→ www.watercooled-adventures.co.uk Tel. 07901 868191
→ www.wye-pursuits.co.uk Tel. 01600 891199

Glamping

Bet you don't know what glamping is. Thought not. Well, it's glamorous camping. Most people would probably say that was an oxymoron. But, in fact, camping has just become chic. A little bit of razz has been put into canvas.

Camping has never been more popular. Over 20 million trips are made to British camping and caravanning parks each year and if that hasn't yet included you then maybe glamping can change your mind. The trend began with the arrival of Mongolian yurts, the circular highly superior dwellings traditionally used by the nomads of east and central Asia. In winter the Mongolian climate can be pretty severe so yurts are accordingly snug, with wood-burning stoves.

The days of arriving in the dark and struggling with a sack of guy ropes and flimsy semi-proofed nylon, only to have the whole lot collapse on you in the night in the rain, have not the remotest connection with glamping. Outside the cows moo and the sheep baa, just like always, but you now arrive to find your abode in place and completely fitted out. You bring no more than you would for a stay in a country hotel.

Glamping sites include:

- **Lake District Yurts, Kendal**. And they're so green they even offer a ten per cent discount to anybody arriving without a car (they'll meet you at the local railway station).
- **Woodland Tipis, Herefordshire**. Tipis are less substantial than yurts but still large enough to have a stove inside. The chimney or smoke hole at the apex has a 'hat' to keep out any rain. To preserve the ambience, cars aren't allowed up to the tipis but you'll be lent a wheelbarrow instead.
- **Featherdown Farm Holidays**. Set up on working farms, these highly superior tents are more like cottages with wooden floors and a flush toilet. At the time of writing there are currently 11 locations from Cornwall to Scotland but more will have been added by the time you read this. The tents have a private master bedroom with a proper bed, are heated by woodburning stoves and lit by oil lamps. In the morning you'll get up and search for fresh eggs, just like the old days.
- **Marthrown of Mabie, near Dumfries, Scotland.** You can sleep here in a tipi but the most stunning accommodation is the replica of an Iron Age roundhouse. You can do your cooking in the middle over an open fire, if you want. The roundhouse sleeps 16 and is perfect, say the owners, for stag or hen parties. Hmm. Who needs cheap flights to Barcelona?
- **Under the Thatch Holidays**. Perched above the beach in Wales's Cardigan Bay you have a choice between meticulously restored Romany caravans, a showman's wagon and even a couple of Edwardian railway carriages.
- **Vintage Vacations.** If you're into retro then these authentically restored vintage Airstream American trailers on the Isle of Wight are just right for you. Shiny aluminium pods, they're as quintessentially 1950s as James Dean.

What's green about this holiday? Camping in Britain is as green as it gets – less impact than a hotel, little or no electricity consumption and close to nature.

Who is it for? If you hate camping you might still enjoy glamping.

What's a green way of getting there? Coach, train or energy-efficient car with all seats taken; if the car isn't full, use Freewheelers to offer the spare places (**www.freewheelers.com**). Bring a bike to use locally.

How much pollution am I saving? By not staying in a hotel you might save around 50kg of carbon dioxide; by camping in the UK rather than flying to, say, the south of France you could save about three-quarters of a tonne.

Anything not green about it? No.

What time of year? Many glamping sites are closed in winter but, with their stoves and fires, yurts and tipis can be comfortable from Easter right through to October.

What price? Tipis come at around £195 weekend/£450 week while yurts and more elaborate dwellings cost a little extra. With several of you we're talking £10-£15 a night with transport and food on top.

Where can I get further information? For the country's most innovative and interesting camping possibilities read *Cool Camping* by Jonathan Knight www.coolcamping.co.uk. For a more traditional approach try www.campingandcaravanningclub.co.uk Tel. 0845 1307631.

How do I make a booking?
➜ www.vintagevacations.co.uk Tel. 07802 758113
➜ www.underthethatch.co.uk Tel. 01239 851410
➜ www.featherdown.co.uk Tel. 01420 86922
➜ www.woodlandtipis.co.uk Tel 01432 840488
➜ www.lake-district-yurts.com Tel. 01539 821278
➜ www.yurt-holidays.co.uk online only
➜ www.larkhilltipis.co.uk Tel. 01559 371581
➜ www.marthrown.com Tel. 01387 247900
➜ www.cornish-tipi-holidays.co.uk Tel. 01208 880781
➜ www.deepdalefarm.co.uk Tel. 01485 210256 (tipis)

Finally, if you'd like to buy a designer tent with a difference (floral fabric, for example) see www.cathkidston.co.uk; prices from about £60, matching wellies and sleeping bag extra.

WHY GO FURTHER NO. 7 – On safari

Shush ... your safari guide motions you to tread softly. He's got a sighting. Get down now. Here it is. The rare, recently re-discovered ... short-necked oil beetle (*Meloe brevicollis*).

Unimpressive he may seem, but in his own way he is a thing of great beauty and, of course, hugely significant to those who realise that this small, black and horned creepy-crawly was thought extinct since 1945. He can now be seen during wildlife safaris organised by the Soar Mill Cove Hotel, set in 10,000 acres of National Trust land near Salcombe, South Devon. Organically managed , the valley does offer larger animals, too, including deer, hares, rabbits and badgers.

But it is in the Scottish Highlands that a safari can hold up its head. To sit in a hide and see otters, golden eagles and red deer, and then to get back to an open fire and a wee dram, is a green dream come true. And let's not forget that a red deer is a pretty impressive animal, with males weighing more than a third of a tonne, while the antlers alone can be 15kg. There are estimated to be 350,000 in the Highlands, a spectacle on an African scale. What's more, the largest creature ever to have lived on the planet – the blue whale – can occasionally be seen in Hebridean waters, while porpoises, dolphins and minke whales are a near certainty.

Here are some suggestions:

- **The Aigas Field Centre, Invernesshire**. This is about as close as you'll get to an African safari in Britain. For a start you can see beavers – the same ones featured on BBC's Autumnwatch. A pair were introduced into the loch in 2006 and subsequently produced the first kit to be born in the Highlands for 450 years. Then there are red and roe deer, otters, badgers, pine martens and a tremendous variety of birds including ospreys, golden eagles, peregrine falcons, honey buzzards, red kites and whooper swans.
- **Speyside Wildlife**. The greatest prize of all, of course, would be a sighting of the Loch Ness Monster. Speyside Wildlife gives you the opportunity on its safari but you'll have a rather better chance of sea eagles, seals and dolphins.

- **Rothiemurchus Estate, Cairngorms**. The quarry here are the beautiful big-eyed Highland cattle, tracked by Land Rover.
- **The Travelling Naturalist, Brockenhurst**. Not all the big mammals by any means are in Scotland. Near this Hampshire village you can also see red and fallow deer. At the other end of the scale, your guide will find spectacular dragonflies.

What's green about this holiday? You're going on safari without leaving the country. And you're giving value to British wildlife.

Who is it for? Everyone who loves wildlife.

What's a green way of getting there? Coach, train or energy-efficient car with all seats taken; if the car isn't full, use Freewheelers to offer the spare places (www.freewheelers.com).

How much pollution am I saving? Compared with flying to, say, South Africa, you could save about seven tonnes of carbon dioxide.

Anything not green about it? It's obviously greener to go on foot than ride around in a Land Rover.

What time of year? Every season has its spectacle. For common seals, for example, it's June/July. For grey seals and red deer it's October/November.

What price? Safaris, even in Britain, aren't cheap because of the expense of hides and guides and so on. For a week, think in terms of £450-£1000.

Where can I get further information?

Put 'wildlife holidays Britain' into your search engine and you'll have plenty of itineraries to choose from.

→ www.travel-quest.co.uk
→ www.bbc.co.uk/nature
→ www.english-nature.org.uk
→ www.wwt.org.uk
→ www.wildlifeextra.com
→ www.wildlifetrust.org

For seasonal ideas try the National Trust www.nationaltrust.org.uk Tel. 0870 4584000 or in Scotland www.nts.org.uk Tel. 0844 4932100

How do I make a booking?

→ www.speysidewildlife.co.uk Tel. 01479 812498
→ www.soarmillcove.co.uk Tel. 01548 561223
→ www.naturalist.co.uk Tel. 01305 267 994
→ www.rothiemurchas.net Tel. 01479 812345
→ www.aigas.co.uk Tel. 01463 782443
→ www.midwalesbirdwatching.co.uk Tel. 01970 890281
→ www.wildlifemull.co.uk
→ www.isleofmullholidays.com
→ www.birdfinders.co.uk Tel. 01258 839 066

Are windfarms green?

Join the John Muir Trust and save Scotland. That's the message. Yes, we want renewable energy. Yes, we want windfarms. But that doesn't mean we want them built in areas of outstanding natural beauty. The John Muir Trust, owner and manager of some of the most wonderful landscapes in Scotland, needs your support to fight windfarm proposals that threaten some of the world's most scenic places. If you enjoy walking in Scotland then take a look at: www.jmt.org

WHY GO FURTHER NO. 8 – Wine tours

Red, rosé or the white? Or maybe a sip of the bubbly? But you're not in the Mosel Valley now. Nor in Burgundy, nor Beaujolais country. And certainly not the Napa Valley. In fact this is the Sedlescombe Organic Vineyard in East Sussex, the longest-established organic vineyard in the country. That makes it about the greenest wine you can get. Not only does it have a far, far lower carbon footprint than imported wine but it's produced with only the minimum of permitted chemicals. Yet Jancis Robinson, Master of Wine no less, describes it as some of the most delicious English wine she has ever tasted.

Over the past 40 years the quality and number of English wines has been growing relentlessly. It's partly thanks to the Gulf Stream, without which growing vines would be almost impossible, and partly due to global warming. The downside of being an island is wet and windy weather which can pose a serious problem during the flowering season.

In fact, there are now so many vineyards throughout the south of England and Wales that you could easily spend a two-week holiday touring round them. Which is exactly what we're suggesting you do.

And, by the way, some have accommodation. Which, when you're tasting wines, is a very sensible idea:

- **Camel Valley Vineyard**, Cornwall. Here you can stay in one of two self-catering cottages. In 2005 the *Cornwall Brut* was the only sparkling wine from outside the Champagne region to win a gold medal at the International Wine Challenge – outpointing 250 Champagnes in the process. If you sign up for the special tour you'll also get a slap-up lunch and lots of different wines to taste. The inland and coastal scenery is stunning.
- **Purbeck Vineyard** is near to the beaches of Swanage, Dorset, a good place to snooze away the post-tasting session. About as old as the new millennium with two very good white wines and a red.
- **Denbies English Vineyard**, Dorking, Surrey, was the 2007 winner of UK Wine Producer of the Year. There's a guest farmhouse on the estate. At harvest time you can do a day's grape picking. But they don't pay you; you pay them £40 – to include lunch and a wine tasting.
- **Hidden Spring Vineyard**, Heathfield Sussex, has some attractive pitches for your tent.

British versus English wine

British wine is not necessarily the same thing at all as English-produced wine. The product calling itself 'British wine', although bottled in the UK, will have been made from concentrated grape juice imported from abroad. These wines tend to be thicker and sweeter than the ones we generally open for dinner these days. An English-produced wine, on the other hand, will come from one of the 400 or so English and Welsh vineyards. Most are members of the United Kingdom Vineyards Association. with its catchy motto 'Think, Drink English'. At their website www.englishwineproducers.com you'll find news, wine festivals and details of vineyards. Also take a look at www.english-wine.com and www.newwavewines.com.

Some facts about English wine

- Although the UK is one of the smallest wine producers in the EU, it is the largest importer (by value) in the world.
- The UK's historical connection with wine goes back to Roman times.
- There are more than 350 registered vineyards in the UK.
- About 3.3 million bottles of wine are produced by English and Welsh vineyards a year.
- England and Wales produces about four times more white wine than red.
- Australia is the largest exporter to the UK followed by France and then the USA.

What's green about this holiday? You'll be touring vineyards without leaving the country.

Who is it for? Even if you don't drink much wine it's a great way of getting out into the countryside – and there are often diversions for the kids, too.

What's a green way of getting there? Bearing in mind you'll probably be buying the odd case or two a car is the practical answer – but it has to be a fuel-efficient model with all seats taken. And, of course, don't taste and drive.

How much pollution am I saving? Compared with flying to, say, Bordeaux you'll be saving at least half a tonne of carbon dioxide.

Anything not green about it? No, as long as you stick to a reasonable mileage.

What time of year? Autumn picking time is probably the most interesting.

What price? Bed and breakfast will cost about £25/£40 per person per night. A double room at Denbies with en suite facilities is £95 a night. Many of the vineyards have special deals combining wine, a wine tour and accommodation.

Where can I get further information?

➜ **www.uk-energy-saving.com**

➜ **www.allorganiclinks.com**

→ www.englishwineweek.co.uk
→ www.thesoilassociation.org.uk
→ www.organicholidays.co.uk

How do I make a booking?

→ www.denbiesvineyard.co.uk Tel. 01306 876616
→ www.camelvalley.com Tel. 01208 77959
→ www.three-choirs-vineyards.co.uk Tel. 01531 890223
→ www.vineyard.uk.com (Purbeck Vineyard) Tel. 01929 481525
→ Adgestone Vineyard, Sandown, Isle of Wight
Tel. 01983 402503
→ www.merseawine.com Tel. 01206 385900
→ Hidden Spring Vineyard, Heathfield, East Sussex
Tel. 01435 812640
→ www.englishorganicwine.co.uk (Sedlescombe)
Tel. 01580 830715 – for accommodation try the nearby organic Slides Farm B&B Tel. 01580 880106

WHY GO FURTHER NO. 9 – Retreats

The dinging of the windchimes and the wind fluttering through the prayer flags are normally sure signs of a retreat high in the Himalayas. But you can get those same externals in Britain, too, while the internal experience stays more or less the same.

Basically there are two kinds of retreats in Britain. Those where you'd be expected to join in. And those where the tranquillity is more of a backdrop and you a sympathetic observer, picking only the activities that appeal to you. Here are some suggestions:

- **Mamaheaven.** These new-mother-and-baby retreats take place at Penrhos Court, near Kington on the Hereford/Welsh border. Combining yoga, individual bodywork, beauty therapies, information on nutrition, alternative remedies and eco baby care they aim to send new mums away feeling a lot more confident and a lot more tranquil. Childcare facilities are available.
- **Heartspring**. A lovely old house at Llansteffan, Carmarthen, with views of the coast. Furnishings and even the paintwork are 'natural'

and spring water comes out of the tap. Therapies include meditation, yoga, massage, reflexology and counselling.

- **Eco Retreats.** Winner of the 2007 Real Alternative Award at the Welsh National Tourism Awards, Eco Retreats combines the best of glamping (tipis and a yurt) with a holistic approach to therapy. It's hidden away in Powys, to the south of Cadair Idris. You could call it 'retreat lite' – Reiki, meditation, nature and the magic of the campfire.
- **Manjushri Kadampa Meditation Centre**. At Coniston Priory, just south of the Lake District in Cumbria, this is a real alternative to a Buddhist experience in the Himalayas. Established in 1975, it offers a variety of courses and retreats, both residential and non-residential. All meals are vegetarian.

What's green about this holiday? You don't have to travel far. You'll probably eat vegetarian meals. You'll be nicer at the end of your stay. And you'll pick up lots of environmental ideas from fellow retreaters – gentle folk, aiming to do no more than commune with their inner selves.

Who is it for? There are many different kinds of retreat for many different kinds of people – including children in some cases.

What's a green way of getting there? Ideally levitation. Otherwise, coach, train or energy-efficient car with all seats taken; if the car isn't full, use Freewheelers to offer the spare places (**www.freewheelers.com**).

How much pollution am I saving? Compared with flying to India you could save about five tonnes of carbon dioxide.

Anything not green about it? No.

What time of year? Any time in warm, cosy retreats – summer for the more austere ones.

What price? The Manjushri Kadampa Meditation Centre charges £175 a week in a single room/£140 each in a double room or £105 in a dormitory, all prices including meals. Other retreats can go as high as £500 for a week.

Where can I get further information? Go to the *Mind, Body and Spirit Festivals* held in London every May and Manchester every October. Details at www.mindbodyspirit.co.uk Tel. 0207 3719191. Other useful websites include the online magazines www.kindredspirit.co.uk and www.resurgence.org, as well as the directories www.retreatsonline.com, www.theretreatcompany.com Tel. 01162 2599211 and www.places-to-be.com.

How do I make a booking?

→ www.mamaheaven.org

→ www.launde.org.uk (Leicestershire) Tel. 01572 717254

→ www.heartspring.co.uk (Carmarthen, Wales) Tel. 01267 241999

→ www.stcuthmans.com (Horsham, West Sussex) Tel. 01403 741220

→ www.samyeling.org (Scotland) Tel. 01387 373323 ext.22

→ www.ecoretreats.co.uk Tel. 01654 781 375

→ www.manjushri.org Tel. 01229 584 029

Down on the farm

Next job is to muck out the pigs. But what's this? The sow seems rather more friendly than you'd like. Take care! Apparently, to a sow on heat, a sweaty human male smells rather similar to a boar – and she'll stop at nothing to get her man! Just one of the essential facts you'll need to know before starting your working holiday on an organic farm. Anything else you should know? Well, never get between a cow and her calf – especially if she's the one with the long horns. Avoid grabbing a goat around the neck – they tend to faint. And don't be tempted to order a shepherd's dog around – they're trained to respond only to their masters' voices and will probably nip you.

If you'd like to experience this world for a time you only have to volunteer. Many farms and vineyards all over the country – very often organic ones – take on volunteer labour in return for full board and lodging. Some of it will involve sweaty jobs like mending fences, digging potatoes and, yes, mucking out the pigs. But you'll also be learning about the science of organics – how farming can be sustainable without recourse to factory methods or the intensive use of chemicals. Useful when you get back home to the vegetable patch. You'll work hard, you'll eat well and – and you'll learn to wash well, too!

If your accommodation is in the hay barn, don't complain. We've tried it and it's great. But, although you won't get five-star accommodation (that's reserved for the paying guests) the digs are usually more than adequate. If it's a bunkhouse or caravan you'll need to bring your own sleeping bag. (Anyway, it doesn't seem to matter so much if you're getting up almost before you've gone to bed, does it?)

If you've never done anything like it before it's probably a good idea to start with a weekend. If you like it you can arrange something longer. In any case, it's a good idea to contact the farmer personally to make absolutely clear what the working holiday will involve and what hours are expected and what board is provided. An average working day might well be eight solid hours with breaks for meals, starting at around 5am. Not for the faint-hearted.

And what are farmers looking for in volunteers? In a survey they put enthusiasm top of the list, followed by good manners, sensitivity to animals, a willingness to get stuck in and a good sense of humour – essential where pigs are involved.

What's green about this holiday? You won't have to travel far, you'll be working on the land and, in the case of an organic farm, you'll be helping to promote a healthier, safer world.

Who is it for? Fit, enthusiastic adults.

What's a green way of getting there? Bicycle, coach, train or energy-efficient car with all seats taken; if the car isn't full, use Freewheelers to offer the spare places (www.freewheelers.com). Some farmers will arrange for a pick-up at the station or bus stop.

How much pollution am I saving? Compared with flying to a holiday in the Med you'll be saving from half a tonne to two tonnes of carbon dioxide.

Anything not green about it? If your farmer is into lots of chemicals and intensive farming you'll have to try to talk him out of it.

What time of year? Any and every but there's a particular need at harvest time.

What price? Just your travel and any extras.

Where can I get further information/make a booking? Track down the book *The Organic & Sustainable Farm Holiday Guide* by Christopher Mager (Two Heads Publishing).World-Wide Opportunities on Organic Farms (www.wwoof.org.uk) specialises in matching volunteers with suitable communes, organic gardens and farms. The Soil Association can provide you with details of organic farmers in your chosen area and you can then contact them yourself – www.soilassociation.org Tel. 01173 145000 or in Scotland www.sascotland.org Tel. 01316 662474. You can get general information about organic food production from the International Federation of Organic Agricultural Movements www.ifoam.org. Farms that often have a need for volunteers include www.obanfarmpark.co.uk Tel 01631 720223; Westbourne Manor Farm, East Sussex www.wowo.co.uk Tel. 01825 723414; Deepdale Farm, Norfolk, www.deepdale.co.uk Tel. 01485 210036; and Sedlescombe Vineyard in Robertsbridge, East Sussex, www.englishorganicwine.co.uk Tel. 01580 830715.

DIRTY WEEKENDS (OR EVEN THE WHOLE WEEK)

We thought that would get your attention. But what we have in mind is the sort of dirt that comes with trying to put the environment back together again. In other words, these holidays don't just minimise the impact on the environment, they actually improve it.

The National Trust is a great place to start. It offers about 500 different types of working holidays each year, with something suitable for every taste and age. There's a terrific spirit of camaraderie and many of the holidays are sold out well in advance to people who come again and again. The Trust looks after some 300 historic houses with gardens, 50-odd industrial monuments and loads of parks and woodlands in England, Wales and Northern Ireland. So there's bound to be something not far from you. Scotland has its own National Trust with similar programmes.

What kind of work might you be doing? It could be anything from gardening to dry stone walling, from hedge-laying to restoring a pond, from coppicing to goat herding. You'll almost certainly pick up new skills from the experts and you'll be able to use them back home. Note that accommodation tends to be spartan.

Other excellent conservation holidays are run by the British Trust for Conservation Volunteers (BTCV) and the Field Studies Council which has 17 centres.

If you're looking for something more specialist, how about a12-day field course run by the Cetacean Research and Rescue Unit (CRRU) in Banff, Scotland in conjunction with the Earthwatch Insitute? Or a five-day liveaboard experience with the Wildlife Trust Holiday team working with basking sharks? With their huge mouths they can seem quite frightening but, in fact, they're harmless. It's guaranteed you'll get wet, exhausted and cold but working with marine mammals is an uplifting experience and a vital task. Incidentally, there are 47 Wildlife Trusts all over the country and although it's already the largest UK voluntary organisation they're always looking for more help for local, weekend projects. Children are welcome.

What's green about this holiday? You don't have to travel far, you'll be improving the environment very directly and you'll be learning useful green skills.

Who is it for? Anybody between the ages of 16 and 70 with a willingness to get dirty fingernails.

What's a green way of getting there? Bicycle, coach, train or energy-efficient car with all seats taken; if the car isn't full, use Freewheelers to offer the spare places (www.freewheelers.com).

How much pollution am I saving? Compared with, say, flying to the Med you'll be saving between half a tonne and two tonnes of carbon dioxide.

Anything not green about it? No.

What time of year? All year and, indeed, Christmas and New Year holidays get sold out very early.

What price? Short-break National Trust or BCVT working holidays cost from as little as £60 all in. Week-long packages will be more like £200 in 'comfortable' accommodation and if you're learning a skill you might have to pay as much as £400. The Wildlife Trust liveaboard trip costs about £600 per person and the CRRU project around £700 for 12 days.

Where can I get further information/make a booking? Put 'volunteering holidays' into your search engine together with the name of your chosen holiday area.

→ www.volunteering.org.uk
→ www.earthwatch.org.uk Tel. 01865 318831
→ www.do-it.org.uk
→ www.cardiff.gov.uk/flatholm Tel. 0292 0353917
→ www.wildaboutbritain.co.uk
→ www.nts.co.uk Tel. 0131 2439300
→ www.nationaltrust.org.uk Tel. 0870 4584000
→ www.field-studies-council.org.uk
→ www.wildlifetrusts.org Tel. 01636 677711
→ www.crru.org.uk Tel. 01261 851696
→ www.bcvt.org.uk Tel. 01302 388888

Down on the farm (Part 2)

There is a whole different way of staying on a farm [see *Part 1* above]. Instead of having to turn out when young Charlie Cockerel wakes everyone up, you just turn over and go back to sleep. In this case, yours is more the role of the gentleman farmer. A leisurely (farmhouse) breakfast, a stroll around to see, stroke and cuddle the animals, maybe a swim in the pond, a leisurely (farmhouse) lunch, a bike ride or a trip to the nearest beach, a leisurely (farmhouse) dinner, then out to watch the bats and the badgers, and so on. And unlike the volunteers, you, as a paying guest, will very probably be enjoying a well-equipped cottage or a beautifully converted barn.

What's green about this holiday? You won't be travelling far, you'll be helping the farmer make a living and you'll be encouraging the preservation of the countryside.

Who is it for? Anyone not too bothered about clubbing.

What's a green way of getting there? Bicycle, coach, train or energy-efficient car with all seats taken; if the car isn't full, use Freewheelers to offer the spare places (www.freewheelers.com). Some farms have bicycles for getting around the local area.

How much pollution am I saving? Compared with flying to the Med, anything from half a tonne of carbon dioxide up to two tonnes.

Anything not green about it? Some farmstays are pretty slick operations.

What time of year? Spring for lambs, summer for the weather and autumn for those lovely rustly days.

What price? Summer full-board from around £350 though possibly less if children are sharing a room.

Where can I get further information?

➔ www.soilassociation.org Tel. 0117 3145000
➔ www.sascotland.org Tel. 0131 6662474
➔ www.ruraldoreset.co.uk
Tel. 01308 422884 (B&B) 01258 820022 (self-catering)
➔ www.farm-holiday-cottages.com Tel. 01983 730783
➔ www.farmstayscotland.co.uk
➔ www.farmstayuk.co.uk
➔ www.visitscotland.com
➔ www.golakes.co.uk Tel. 0845 4501199
➔ www.walestouristonline.co.uk
➔ www.stayinwales.co.uk

How do I make a booking?

Organic farms taking guests include:

➔ www.beechenhill.co.uk Tel. 01335 310274 (Ashbourne, Derbyshire)
➔ Blaeny Nant (no web) Tel. 01248 600400 (Gwynedd, Wales)
➔ www.clynfyw.co.uk Tel. 01239 841236 (Pembrokeshire, Wales)
➔ www.hindonfarm.co.uk Tel. 01643 705244 (near Minehead, Somerset)
➔ www.littlecomfortfarm.co.uk Tel. 01271 812414 (North Devon)
➔ www.penyraltt.co.uk Tel. 01559 370341 (Carmarthenshire)
➔ www.slackhousefarm.co.uk Tel. 01697 747351 (Cumbria)
➔ www.westhillfarm.org Tel. 01271 815477 (Ilfracombe, Devon)

Other farm stay holidays can be booked through:

➔ www.featherdown.co.uk Tel. 01420 80804
➔ www.poyerstonfarm.co.uk Tel. 01646 651347
➔ www.tyddynsyrhuwfarmholidays.co.uk Tel. 01286 650679
➔ www.glascoedfarm.co.uk Tel. 01994 231260

WHY GO FURTHER NO. 10 – Foreign food

What do you go abroad for? Lots of things. But food and wine probably come pretty high on most lists. And yet the irony is that Britain has probably the most cosmopolitan cooking in the world. French, Italian, Greek, Indian, Thai, Chinese, Mexican, Caribbean ... The list goes on and on. And when it comes to wine, you'll find an infinitely greater range of choice in a supermarket in Tunbridge Wells or a restaurant in Southampton than you'll ever get in France. Nor should we forget British cooking itself. It has an inferiority complex but it's better than it thinks. Whether we're talking restaurants or whether we're talking ingredients for home cooking, Britain is about the most sophisticated country on the planet.

So the message is, don't go abroad for food. Stay here and learn to cook all those dishes for yourself. And to make it really green we're suggesting you concentrate on the vegetarian and vegan elements from various national cuisines. (Raising animals for meat, don't forget, results in a reduction in available protein, wastes huge quantities of water and increases the amount of carbon dioxide and methane in the atmosphere.)

- **Green Cuisine** is a pretty good place to start. In a converted barn on the Welsh borders it offers two or five-day packages which not only focus on vegetarian, organic ingredients but also include yoga and massage. You'll learn things like 'the importance of the right fats in our system', how to sprout seeds for the kitchen, and the creation of dishes such as 'vegetable cannelloni with smoky pepper sauce' and 'marinaded tofu with coconut and lemongrass rice'.
- **Meadow Barn Cookery School**, Herefordshire. Here you can learn all about bread, pastry, pasta, vegetables and fruit with an emphasis on locally sourced, seasonal ingredients.
- **Bread Matters** at the Village Bakery in Melmerby, Cumbria, is the place to get down to the basic foodstuff in so many different countries. Here you'll learn the difference between *Altamura semolina* and *Tuscan schiacciata*, using mostly organic ingredients. Under the watchful eyes of tutors 'students learn and understand what is really

happening when flour, water, salt and yeast are mixed together'. Lunch is included, as is one (organic) restaurant dinner and you can choose from a list of local accommodation. There are also three- and five-day courses if you want to move on to the alchemy which is sourdough baking.

- **Gables School of Cookery**, Gloucestershire. Another opportunity to improve your international vegetarian skills.

What's green about this holiday? You'll be cooking and eating foreign food without travelling outside the country. And, hopefully, it will be vegan or vegetarian, too.

Who is it for? Adults of both sexes and (usually) children over 14.

What's a green way of getting there? Bicycle, coach, train or energy-efficient car with all seats taken; if the car isn't full, use Freewheelers to offer the spare places (www.freewheelers.com).

How much pollution am I saving? By not flying to Lyons to sample the gastronomy of France you'll be saving at least half a tonne of carbon dioxide.

Anything not green about it? No.

What time of year? Any time.

What price? Two-day bread course £375; three-day £495 and five-day £650, plus modest accommodation expenses. Meadow Barn's weekend breaks are around £400, more or less full-board (put it this way, you won't have to go out to find anything more to eat). Green Cuisine asks around £275 for two days or £699 full board for five days, including the yoga and massage.

Where can I get further information/make a booking?

→ www.breadmatters.com Tel. 01768 881899

→ www.cookawayfoodbreaks.co.uk Tel. 01951 580249

→ www.thegablesschoolofcookery.co.uk Tel. 01454 260444

→ www.greencuisine.org Tel. 01544 230720

→ www.stirringstuff.co.uk Tel. 01491 628523

→ www.eckingtonmanorcookeryschool.co.uk Tel. 01386 751362

Organic really is better

There's been a lot of controversy about organic food. There are always those who want to knock it, along with any effort to save the environment or make the world a better place. Well, now it's more or less official that organic really is better. A European Union-funded study has shown that there are up to 40 per cent more antioxidants in organic fruit and vegetables than in industrial versions. Antioxidants are substances that combat cancer, heart attacks, strokes and many age-related diseases. Organic produce also contains more vital minerals such as iron and zinc. And when it comes to milk, the antioxidant content of organic is 90 per cent higher than non-organic.

- Organic carrots, potatoes and kiwi fruit have more vitamin C
- Organic milk has higher levels of vitamin E
- Organic spinach, lettuce and cabbage have more minerals

WHY GO FURTHER NO. 11 – wintersports

When it comes to winters, Britain is pretty good at them. So why bother to go anywhere else for wintersports?

OK, admittedly there can be a problem of actual snow. British mountains in winter are cold enough but they're not always covered by the white magic. Incredible – given the water all around us – but true. Yet, if the forecasts of wetter winters are correct Britain could become a wintersports paradise.

In any event, the thing to do is *not* to book ahead. Try to arrange things so you can be flexible and take your wintersports holiday once the snow reports are favourable. Then head for Scotland. Just think. No airport queues. No channel to get across. No rush. If you live in, say, the Manchester area you can leave mid-morning and be there in time for dinner – just 300 miles or so (around 500km).

These are Scotland's resorts:

- **Cairngorm** – the most famous and based on the resort of Aviemore. It has 16 lifts and 23 miles (37km) of runs.

- **Nevis Range** – opened in 1989, this is Scotland's highest resort, with runs on the slopes of Aonach Mor, Britain's eighth-highest mountain. Fort William is just 10 minutes away by car or shuttle bus. There are 11 lifts and 22 miles (35km) of runs.
- **Glenshee** – to the south of Cairngorm, Glenshee has 23 lifts and 25 miles (40km) of runs.
- **The Lecht** – to the north of Cairngorm, not far from Tomintoul, this is a good beginner's area. It has plenty of artificial snowmaking so you should be assured of something to slide about on. The figures are 12 lifts and 12 miles (20km) of runs, although all of them are short.
- **Glencoe** – over towards the west coast, just to the south of Fort William, Glencoe has seven lifts and 12 miles (20km) of runs, making it rather more limited than the top three areas, but it does have the longest descent. If you're based at Fort William it makes a good change of scene when you're tired of the Nevis Range.

If you've already skied in the Alps or the Pyrenees you may find Scotland a touch colder and less sunny. But if the snow comes late then March or even April can be excellent months. Just make sure you've got good clothing.

By not going abroad to ski you're already being pretty green. But you'll be even greener if you switch to the cross-country style of skiing. It's not just that ski lifts use energy. They're also extremely ugly. In winter, when there's snow on the ground, they're somehow more acceptable. But come the spring they disfigure the wilderness.

A word of warning

The Cairngorm area is a good place to get started but, at the same time, a lot of the uplands are very fragile. So don't just go charging about anywhere. First of all, get a little bit of advice locally about where it's acceptable to go and where it isn't.

DOG POWER

You don't have to go to the Yukon for dog mushing. You can try it in Scotland with Alan and Fiona Stewart, near Aviemore, where they have about 30 sled dogs. Just being with all those boisterous, excited animals is an experience in itself. But it gets even better on the night run, with a good chance of seeing red deer. You'll also get the opportunity to drive a team yourself.

What's green about this holiday? It's a domestic alternative to travelling to the Alps, the Pyrenees or North America.

Who is it for? Beginner to intermediate downhill skiers, all cross-country skiers and anyone who fancies a ceilidh in the evenings.

What's a green way of getting there? Train, coach or car with all seats taken.

How much pollution am I saving? Compared with flying to the Alps you could save around 350kg of carbon dioxide; compared with flying to North America you'd be talking about a saving of more like four tonnes.

Anything not green about it? Ski lifts do disfigure the mountains – cross-country skiing is preferable.

What time of year? Good snow tends to come late in Scotland (say March or April) – you need to keep a close watch on the snow reports.

What price? Think in terms of around £250 for a week half-board, plus the cost of getting there.

Where can I get further information? Take a look at www.ski.visit scotland.com or Tel. 0845 22 55 121. For dog mushing see www.sled-dogs. co.uk or www.rothiemurchus.net.

How do I make a booking? The above website lists various ski operators. For dog mushing telephone 07767 270526 or e-mail Stewartsledk9@ aol.com.

Chapter 9

Green holidays on the island of Ireland

A lot of people probably associate Ireland, especially the north, far more with *The Troubles* than with holidays. So we're going to start off with a very special peace experience. But, first of all, if you don't already live there, what's the most environmental way of arriving?

Both planes and ferries generate far more pollution than land transport [see *Chapter 1*] so the greenest way is the shortest sea route or the shortest flight. The ideal would be to hike down the Kintyre peninsula in Scotland and swim across the North Channel – a feat roughly equal to swimming the English Channel. But for all those who are less than superhuman the preferred choices are:

- Stena Lines ferry from Stranraer (SW Scotland) to Larne (25 minutes from Belfast) or Belfast itself.
- P&O Irish Ferries from Cairnryan (five miles north of Stranraer) to Larne.

In the next category, in descending order of preference, come:

- Stena Lines ferry from Fishguard (in Pembrokeshire) to Rosslare (County Wexford on the SE tip of Ireland) or Irish Ferries from Pembroke to Rosslare.
- Irish Ferries from Holyhead on the island of Anglesey, North Wales, to Dublin City (on what's claimed to be the largest ferry in the world); or Stena Lines ferry from Holyhead to Dun Laoghaire (a seaside resort seven miles south of Dublin with good public transport links).

- Flight from Blackpool, Glasgow or Liverpool to Belfast.
- Flight from Blackpool or Liverpool to Dublin.

Ferry details from:

→ www.stenalineferries.co.uk Tel. 0870 707070
→ www.irishferries.com Tel. 0870 7171717
→ www.poirishsea.com Tel. 0870 2424777
→ Flight details from www.flightmapping.com

In other words, it's better to travel *further* by coach, train or fuel-efficient car with all seats taken, in order to take a *shorter* trip by ferry or plane. (To offset carbon emissions for your ferry journey see www.ferrygreen.com Tel. 0870 2642644.)

The peace tour

From every point of view, war and terrorism are the very opposite of green. Anything that can be done to affirm the peace process, promote understanding and generate employment through tourism is environmental. So when you're in Belfast be sure to take the Coiste Political Tour in the Gaeltacht Quarter.

Led daily (11am, Sundays 2pm) by a former political prisoner, the two-hour tour from the Falls Road takes a *real* look at the local community from an insider's point of view. We suggest you combine it with two other itineraries, the so-called Belfast Safari and the Pub Tour. In the former you knock on a door and get a nice cup of tea; in the latter you fall out the door.

What's green about these tours? They help give value to peace.

Who are they for? Adults who want to understand better.

What's a green way of getting there? The tours are on foot; for the greenest ways of getting to Belfast see above.

What time of year? There are tours all year.

Where can I get further information/make a booking?
www.gotobelfast.com

www.belfastcitycouncil.gov.uk Tel. 02890 320202
www.coiste.ie/politicaltours Tel. 02890 200770
www.belfastsafaris.com Tel. 02890 222925
www.belfastpubtours.com Tel. 02890 330844

WHY GO FURTHER NO. 12 – Great wonders of the world

If you were asked to name some of the great, natural wonders of the world you'd probably mention, say, Victoria Falls, the Grand Canyon and the Great Barrier Reef. But a genuine wonder and UNESCO World Heritage Site exists just 50 miles from Belfast, on the northern tip of Ireland. Indeed, in the 18th century it was known as the Eighth Wonder of the World. This extraordinary place is the Giant's Causeway, a staggering natural pavement of 40,000 regular, tightly packed hexagonal basalt columns, each large enough to stand on.

For many years after the Bishop of Derry brought word of the miraculous causeway following a visit in 1692, people debated how so many perfect columns could possibly have come about. One theory was that they were carved by hand by early people. Another was that they were the work of the giant Finn MacCool. But modern science has explained how they're the result of the crystallisation of hot magma on coming into contact with the sea. You can climb about on them as much as you like – at 63 million years old or so they're not going to get worn away very easily. One other astonishing fact – at that time the region was close to the equator.

The causeway and its setting under the spectacular cliffs of County Antrim – recently voted the world's fifth most important view – are reason enough to come but if you need another, the nearby village of Bushmills is famous for its whiskey (not whisky).

What's green about this holiday? You're remaining in the UK and giving value to the preservation of a World Heritage Site.

Who is it for? Everybody.

What's a green way of getting there? Train from Belfast to Portrush and then by the regular bus service (details at www.nirailways.co.uk). Or you could cycle the 50 miles.

How much pollution am I saving? By not flying to other basalt formations such as Castellfollit de la Roca in Catalonia, Garni Gorge in Armenia, the Devil's Tower in Wyoming, the Devil's Postpile in California or the Organ Pipes on New Zealand's Mount Cargill, you'll be saving between half a tonne and 11 tonnes of carbon dioxide.

Anything not green about it? No.

What time of year? Any time.

What price? The Giant's Causeway belongs to the National Trust and is free.

Where can I get further information?
➜ www.northantrim.com/giantscauseway.htm
➜ www.discovernorthernireland.com Tel. 02890 0231221
➜ http://whc.unesco.org
➜ www.giantscausewaycentre.co.uk Tel. 02820 731855
➜ www.nationaltrust.org.uk

How do I make a booking?
➜ www.minicoachni.com Tel. 02890 315333
➜ www.irondonkey.com Tel. 02890 813200
➜ www.killary.com Tel. 00 353 9542276
➜ www.pedalpowercycleireland.co.uk Tel. 02890 715000
➜ www.railtours.ie Tel. 00 353 18560045
➜ www.bestvaluetours.co.uk Tel. 0870 2256263
➜ www.giants-causeway-hotel.com Tel. 02820 731210
➜ www.allthingseco.co.uk (green accommodation directory)

WHY GO FURTHER NO. 13 – The pilgrimage

Think of pilgrimages and you probably think of Santiago de Compostela in Spain, of Rome or even of India. But Ireland has plenty of sites of its own. And several of them have been parcelled together by eco-tourism operator Greenbox, embracing all faiths and none. Focused on the 700-year-old pilgrim site of Station Island in north-west Ireland's Lough Derg (it means 'red lake' in Irish), the 14-day package is a trek on foot promising contrasts of excitement, fascination, education, fun and, well, downright discomfort actually. It starts (and ends) at the Gyreum, described variously as Ireland's 'quirkiest' eco-lodge, 'Christmas pudding-shaped' or 'Hobbit-house', and progresses via hospitality from Knights Templars, Celtic Hermits, Anglican Cub Scouts and Tibetan Buddhists. St Patrick's 'Purgatory' on Station Island is the highlight where you will face a two-day sleepless fast, just, it is said, as St Patrick himself did. There will, we are promised, also be plenty of feasting. For a flavour of the place read Irish poet Seamus Heaney's *Station Island* collection.

What's green about this holiday? It's on foot, it's inter-faith – so promoting understanding between different beliefs – and the accommodation is pretty ecological.

Who is it for? The dedicated and the curious.

What's a green way of getting there? Coach from Belfast or fuel-efficient car with all seats taken. Lough Derg is about 150 miles west of Belfast, just across the border.

How much pollution am I saving? By not flying to Rome you'll be saving about a tonne of carbon dioxide.

Anything not green about it? No – as long as you don't fly very far.

What time of year? Usually March

What price? Around £1,100 fully inclusive.

Where can I get further information?

➜ www.ireland.com

➜ www.gyreum.com Tel. 00 353 719165994

How do I make a booking?

www.greenbox.ie Tel. 00 353 719165994

Northern Ireland's island

Northern Ireland's largest seabird colony is on Rathlin Island. And it's not hard to see why. With sheer sea cliffs reaching as high as 1,500ft (470 metres) and a population of just 80 on a wild wind-blown island measuring eight miles by one, there's very little to disturb them. At the

West Light Viewpoint you can look along the dramatic cliffs and sea stacks and watch kittiwakes, fulmars, guillemots, puffins and razorbills wheeling, crying and nesting in their thousands.

You could make it a daytrip from Ballycastle but for the almost unique experience of being somewhere without a car (only a few locals have them) you need to stay a while. That's not so easy because there's only one proper hotel (the Manor House Guesthouse), one B&B, one hostel and one sort of refuge near the Kebble National Nature Reserve. So book well in advance. By UK standards it doesn't get much more remote than this.

Not only the birds take refuge here. According to legend, a cave in the cliffs is where Robert the Bruce famously watched a spider painstakingly rebuilding its web and resolved to make yet one more attempt to win the throne of Scotland. He succeeded and became Robert I of Scotland from 1306-1329. Fascinating as spiders are, the scenery along the coast is even better, forming part of the Antrim Coast and Glens Area of Outstanding Natural Beauty.

What's green about this holiday? You'll be supporting the conservation efforts of the Royal Society for the Protection of Birds.

Who is it for? Bird lovers and families who love the outdoors.

What's a green way of getting there? Coach from Belfast to Ballycastle, or fuel-efficient car with all seats taken. Then by Caledonian Macbrayne ferry for the six-mile sea trip.

How much pollution am I saving? By not flying to a Mediterranean island you could be saving around three-quarters of a tonne of carbon dioxide.

Anything not green about it? No.

What time of year? Best time of year to see the birds is May and June.

What price? The ferry trip is £8.60 return. Accommodation at all prices.

Where can I get further information?

➜ Royal Society for the Protection of Birds www.rspb.org.uk
Tel. 01767 693690

➜ www.northantrim.com/rathlin_island.htm

→ www.calmac.co.uk Tel. 08705 650000

→ www.nationaltrust.org.uk

How do I make a booking?

→ Caledonian MacBrayne 028 20769299

→ www.donard.com/cottage/rathlin.asp Tel. 0289 1659

→ www.rathlinmanorhouse.co.uk Tel. 02820 763964

→ Access to West Light Viewpoint by arrangement with the RSPB Tel. 028 207 63948

The greenbox

The greenbox is the name given to what's being described as Ireland's first genuine ecotourism destination. It's a cross-border area in the north and includes the counties of Fermanagh, Leitrim, West Cavan, North Sligo, South Donegal and North West Monaghan. The landscape is wild and fabulous and embraces the coastline, too, around Donegal Bay and the beautiful Inishmurray Island. For lovers of outdoor activities, it's one of the greatest places on Earth.

→ www.greenbox.ie Tel. 00 353 719856898

AND THE GREEN JERSEY

Traditionally, the leader in a cycle race wears a yellow jersey. But we're suggesting you put on a green jersey to show you're more interested in the environment than winning anything. Then sign up for the Lap the Lough charity bike marathon. The lough in question is Lough Neagh, the largest freshwater lake in the UK, just 20 miles (32km) west of Belfast and measuring 20 miles (32km) long by nine miles (14km) wide. Your job is to try to cycle right round it on the last Sunday in August. Don't worry. It's nothing like 100 miles. It's only, er, 95 miles, actually (150km).

The thing is, you don't have to finish. Thousands of people will be taking part and although some of them are yellow jersey types, most will just be tootling along the waymarked Loughshore Trail as far as they can manage, enjoying the scenery, the historical sites, the restaurants and, no doubt, the pubs, too. It's a great day out.

And if you don't actually fancy cycling the lough you can nevertheless still enjoy the whole spectacle. The best way is to take one of the boat trips. A favourite is to the non-island known as Church Island (which, theoretically, you could also reach by wading through a marsh) to see the non-church known as Hervey's Folly – just a fake spire with nothing underneath it.

But if you'd like to cycle for one or two weeks rather than just one day, we suggest the gentle and beautiful south-east Ireland. You'll follow the undulating lanes of the Barrow, Nore and Suir valleys, jaunting along at your own pace from one pre-booked B&B to the next, while your luggage is transported by the organisers, Celtic Cycling. The company is based at Lorum Old Rectory, nestling below the Blackstairs Mountains, where you can begin and end your tour.

What's green about this holiday? Cycling is the most environmental form of wheeled transport.

Who is it for? Anyone who enjoys physical activity.

What's a green way of getting there? For Lough Neagh take the bus from Belfast to Lurgan and from Lurgan (Market Street) to Kinnego Marina. Cycles are carried on the bus if there's room (www.translink.co.uk Tel. 02890 666630). Or arrange a lift via www.rurallift.ie. For Celtic Cycling take the train from Dublin Heuston, or the Expressway bus from Dublin Busaras, destination Muine Bheag (the Irish name for Bagenalstown).

How much pollution am I saving? Cycling is non-polluting. Compared with flying to a Mediterranean holiday you could save between half a tonne and two tonnes of carbon dioxide.

Anything not green about it? No, as long as you don't fly very far.

What time of year? Summer

What price? Entry to Lap the Lough is about £25 and the money goes to charity. One week with Celtic Cycling costs around £525 B&B; two weeks £850.

Where can I get further information/make a booking? For Lap the Lough www.lapthelough.org Tel. 07717 353268 and www.loughshoretrail.com. For cycling holidays taking in the lough see www.irondonkey.com, www.cti-ni.com and www.ballycanalmanor.com. For general information on the areas around the lough (five of the six counties of Northern Ireland have shorelines around them) www.discovernorthernireland.com Tel. 02890 0231221 and www.kingdomofdown.com. For Celtic Cycling www.celticcycling.com Tel. 00 353 98 66650.

IRELAND BY PONY

The English love dogs but the Irish love horses. And what better way to enjoy that emerald scenery than on an animal that can eat it?

You surely know that song *Where the mountains o' Mourne sweep down to the sea*. It's about as instantly Irish as *Danny Boy*. The mountains referred to are in County Down, Northern Ireland, about 40 minutes west of Belfast, rising to the province's highest summit – Slieve Donard at 2,700 ft (850metres). And they really do sweep down to the sea, at the east coast resort of Newcastle. There you'll find the Mourne Trail Riding Centre, from where you can set out through the enchanting Tullymore Forest Park, along the beaches or, indeed, up into the foothills of the Mourne. As you'd expect, you'll be aboard Irish hunters from 15 hands to 17 hands (and if you don't know what that means then this particular stables isn't for you).

You'll also need to be a fairly experienced rider if you want to tackle the famous Ring of Kerry in the musically named Macgillycuddy's Reeks, Ireland's highest mountain range. But if you are, this is some of the finest riding in Europe and as unforgettable as the name. On the one-week itinerary you'll walk, trot and canter from guest house to guest house through fields, bogs, forests and, of course, beaches. A gallop along the four miles of golden sands at Rossbeigh will probably be the highlight.

But not everyone who comes to Ireland is an expert rider. If you've never done it before then the County Offaly Equestrian Centre, about one hour from Dublin, could be the place for you. The accommodation is extremely comfortable, you can learn at your own pace, and when you're not riding there's wonderful hiking and cycling in the Slieve Bloom Mountains.

What's green about this holiday? You don't have to travel far to get there; horse riding is one of the greenest ways of enjoying the landscape.

Who is it for? If you choose your stable carefully most fit people can find a programme to suit (many riding holidays do have provisions for non-riding family or friends).

What's a green way of getting there? For the Mourne, bus from Belfast. For Macgillycuddy's Reeks, train from Dublin to Killarney. For County Offaly Equestrian Centre, train from Dublin to Tullamore.

How much pollution am I saving? By not riding in the USA, the home of the cowboy, you'll be saving at least five tonnes of carbon dioxide.

Anything not green about it? Horses produce plenty of the greenhouse gas methane.

What time of year? Spring to autumn.

What price? A two-hour ride in the Mourne costs about £25; riding Macgillycuddy's Reeks costs about £750 for seven days; a week at County Offaly costs about £600.

Where can I get further information?

→ www.mournemountains.com
→ www.discovernorthernireland.com
→ For accommodation generally see www.holidayhomes-ireland.com
→ For accommodation offering views of the Mourne mountains and the sea try www.thebriers.co.uk Tel. 02843 724347 or www.mournecottages.com Tel. 02843 751251.
→ www.claretourismcouncil.ie Tel. 00 353 656 8288366
→ www.galway.net Tel. 00 353 91395050
→ www.horsesport.ie
→ www.aire.ie Tel. 00 353 45431584
→ www.ehi.ie

How do I make a booking?

→ www.mournetrailridingcentre.com Tel. 028 4372 4351
→ www.hiddentrails.com Tel. 0207 8702282
→ www.irishhorseriding.com Tel. 00 353 61927411
→ www.horse-holiday-farm.com Tel. 00 353 719166152
→ www.equitour.co.uk Tel. 0800 043 7942

Seaweed baths

The idea of bathing in seaweed may not sound very appealing but lots of Irish people swear by it. First seaweed (*Fucus serratus*) is steam-treated and then added to the bath water. The minerals are said to be extremely beneficial for anyone suffering from arthritis and rheumatism. Some of the spas are worth visiting simply for their Edwardian-style architecture.

Contacts:
www.celticseaweedbaths.com Tel. 00 353 719168686; or www.bestloved.com Tel. 00 353 9636238 (both County Sligo).

Pick the flower label

The EU has a scheme for recognising accommodation that meets the highest ecological standards. Not many have so far risen to its stringent requirements but in rural Ireland there are a few. When you're going around or looking at advertisements you'll know them by the flower symbol. Here, so to speak, is the pick of the bunch.

Arch House (Farm Guest House), Tullyhona, County Fermanagh, Northern Ireland http://archhouse.com Tel. 02866 348452.

Clancy's of Glenfarne (B&B and restaurant), Glenfarne, County Leitrim (No www). Tel. 00 353 719853116

Corralea Activity Centre (self-catering cottages), Belcoo, County Fermanagh, Northern Ireland. www.activityireland.com Tel. 02866 386123

Creevy Cottages, Ballyshannon, County Donegal www.creevyexperience.com Tel. 00 353 719852896

Hill View Lodge (B&B), Boyle, County Sligo (No www) Tel. 00 353 191654290

Littlecrom Cottages (self-catering), Newton Butler, County Fermanagh. www.littlecromcottages.com Tel. 02867 738074

Old Rectory (B&B), Ballinamore, County Leitrim www.theoldrectoryireland.com Tel. 00 353 719644089

Prospect Bay, Ballyconnell, County Cavan (No www)
Tel. 00 353 499523930

For information about the scheme generally take a look at www.ecolabel-tourism.eu and www.eco-label.com.

A BOG STANDARD HOLIDAY

You probably don't like the sound of a bog standard holiday. But then, bogs are very underrated places. For a start, they're rich in carbon, removed from the atmosphere by plants over thousands of years. So any peat cutting or disturbance can release significant quantities of carbon dioxide as well as methane into the atmosphere. They may not look much but bogs are vital.

Which is where you come in. According to the Irish Peatland Conservation Council (IPCC), the country is 'the global headquarters' for blanket bogs, possessing, as it does, eight per cent of Europe's entire stock of these wetlands. (Scotland comes second with five per cent.) And the Council intends it to stay that way, for the good of us all. So we're suggesting you volunteer to protect some Irish bogland. You'll be over your welly-tops in it at times, pulling out recalcitrant and invasive pine and birch saplings or participating in one of the scientific surveys. But when you've finished you'll have done a great thing for the planet.

And you don't have to commit to more than you can manage. Based at the Bog of Allen Nature Centre in Lullymore, County Kildare, you can do as many days as you fancy.

What's green about this holiday? The world's peatlands hold two thousand million tonnes of carbon, equivalent to 70 years of global emissions at current levels. So you're doing your bit by keeping the stuff locked in.

Who is it for? Anyone willing to muck in.

What's a green way of getting there? The no.120 bus from Dublin to Allenwood. If you're going to opt to walk the final three miles from the bus stop to the Bog of Allen Nature Centre then ask the driver to drop you at Skew Bridge.

How much pollution am I saving? It's hard to quantify but it could be vital.

Anything not green about it? No.

What time of year? Spring to autumn.

What price? You can slosh away for free. Just budget for living expenses and maybe a pair of higher wellies.

Where can I get further information?

→ www.ipcc.ie Tel. 00 353 45860133
→ www.wwt.org.uk (Wildfowl and Wetlands Trust)
→ www.wetlands.org
→ www.ramsar.org
→ www.nationaltrust.org.uk Tel. 0870 2403207
→ www.cvi.ie Tel. 00 353 18905625 (Conservation Volunteers Ireland)
→ www.wwoof.net??

How do I make a booking?

With the IPCC (see above). For accommodation locally try www.robertstownholidayvillage.com Tel. 00 353 45870870 (self-catering cottages with 'big' showers which you'll need after this one). More accommodation details from http://kildare.ie

TRADITIONAL GAELIC SPORTS

Don't leave Ireland without seeing some traditional Gaelic sports because, let's face it, you're never going to get the chance to see hurling or camogie in Tunbridge Wells. Hurling is quite like hockey or lacrosse and played with a curved wooden stick called a hurley and a hard ball called a sliothar. But it is also a bit like rugby – the bit that calls for muscle power, brute strength and lot of grunting. Camogie is the women's version of the game and the Irish have been playing them both for about 2,000 years. The games would often last all day or until somebody said something like "Whoa there, lets talk about it ..." But now there are two 30-minute halves.

The generally accepted best place to watch top-notch Gaelic sports is in Dublin's Croke Park Stadium www.crokepark.ie Tel. 00 353 18192300. Just 15 minutes from the city centre and easily accessible by bus, there are championships during the summer but tickets are scarce (try www.ticketmaster.ie). If you can't make the stadium, find a local sports bar and watch a match there – the atmosphere can be pretty authentic www.sports-in-bars.ie. For more information try www.gaa.ie.

General information for travel to Ireland

Useful general websites:

- → www.visitnorthernireland.com Tel. 02890 0231221
- → www.discoverireland.ie Tel. 0800 039 7000
- → www.irishtourist.com Tel. 0845 0264621
- → www.lookintoireland.com
- → www.greenbox.ie Tel. 0871 9987099

Useful transport websites:

- → www.nationalrail.co.uk Tel. 0845 7484950
- → www.stenaline.co.uk Tel. 0870 5455455
- → www.translink.co.uk or www.nirailways.co.uk
  Tel. 02890 666630. www.sailrail.co.uk .Tel. 08450 755755
- → www.irishrail.ie. Tel. 00 353 1850 366222
- → www.seat61.com
- → www.rurallift.ie Tel. 00 353 7196 43933 – lift-sharing in rural areas.

Useful accommodation websites:

- → www.allthingseco.co.uk
- → www.yha.org.uk
- → www.farmstaynorthernireland.co.uk
- → www.farmstays.ie

Chapter 10

Green holidays in mainland Europe

It's only a few miles away but, nevertheless, it's 'abroad'. And you know it as soon as you arrive. The language is different, the food is different and the culture is subtly different, too. Nowadays, though, you're unlikely to see men in striped jerseys and berets riding round on bicycles with strings of onions round their necks. However, riding around France on a bicycle is exactly what we're proposing *you* should do. *Zoot, zeese forainers are sooo funneee.*

Cycling holidays

You were probably expecting to read about cycling holidays and we're certainly not going to disappoint you. But don't forget that the key to a green cycling holiday is how you get to the destination. Cycling isn't green if you fly to your start point. And for France there's really no need. You can so easily take your own bike on the European Bike Express, for example (see *Part One*), you can put it on your car (provided it's fuel efficient and with most seats taken), or you can travel by train and pick up a bike when you get there. But you can't fly.

If you weren't considering a cycling holiday we're going to try to persuade you. Or rather, we're going to let Susi Madron persuade you. Something like 25 years ago she went cycling in France with her family and enjoyed it so much she decided to start her own holiday company. The part she liked was meandering through the countryside to a nice little restaurant or picnic spot and then meandering back again (possibly in a slightly more wobbly manner) for a nice hot bath in a comfortable, country hotel.

And thus *Cycling for Softies* was born. We actually had a short break with *Cycling for Softies* a few years back and very pleasant it was, too. Quite a few other companies have come up with the same sort of idea but *Cycling for Softies* remains the market leader in France. We sampled Burgundy (the area and also the wine) but you can also do Alsace, Mayenne, The Loire, Venise Verte, Cognac, the Dordogne, Gascony and Tarn.

The great thing about these holidays is that you just do your own thing. From a list of hotels and restaurants, all personally inspected by Susi, you make your selection and the company arranges everything for you. Including, of course, dealing with any bike problems – no need to get greasy hands. For example, you could base yourself at one hotel and make daily excursions into the surrounding countryside, each day sampling a different and very delightful restaurant. Think in terms of swishing along quiet lanes fringed by plane trees, picnics in fields full of poppies, comfortable family-run hotels and lazy, gastronomic dinners. On the other hand, if you want to make it a preparation for the Tour de France that's up to you. It's your choice. In which case, you won't risk putting on any weight from the four-course meals.

What's green about this holiday? You're touring without causing pollution.

Who is it for? Everybody. Child seats (for kids 3–5 years) and trail-a-long bikes (kids 5–8 years) are available on request.

What's a green way of getting there? Eurolines coaches will take you to most *Cycling for Softies* destinations, as will the train; otherwise you could hitch with Freewheelers (**www.freewheelers.com**) or go in your own car (via Eurotunnel) – if the car isn't full, use Freewheelers to offer the spare places.

How much pollution am I saving? If you take the train to, say, Burgundy, you could save around 300kg of carbon dioxide compared with flying, plus another dollop by not actually using a car during your holiday.

Anything not green about it? No.

What time of year? Mostly summer.

What price? For a week in high season think in terms of £1,000 for seven nights including accommodation, breakfast, a four-course dinner, bike hire, information pack and the services of the local representative.

Where can I get further information/make a booking? www.cycling-for-softies.co.uk Tel. 0161 248 8282.

*

Next let us tell you about the ultimate cycling holiday for people who don't think they can cycle just a few miles on the flat. Even if you're completely out of condition you can do this. Because the company, *Freewheel Holidays*, had the very cunning idea of taking you to the top of a mountain, more or less, and letting you meander down again over the course of a few days. The average distance each day is a mere 20 miles or so and routes are chosen not only for their downhill trajectory but also for tranquillity and an easy-riding surface. To find suitably long, gentle slopes you'll have to travel a bit further, but it's still perfectly convenient without flying.

Apart from pedalling, the other worry for easy riders is luggage. How do you get a suitcase on a bicycle? With an organised holiday you don't have that problem. Freewheel, like most operators, will transport your luggage to the next hotel, leaving you unencumbered.

Take their grandly entitled *Into the Valley of the Alps*, for example. They transport you to the Austrian village of Krimml, close to Europe's highest waterfalls (1250ft), and after a day to relax and get your bearings they help you onto a bicycle, give you a gentle push and off you go. Once gravity has guided your bicycle the 20-odd miles to Mittersill you slide off, massage your buttocks, and prepare to enjoy another beautiful Austrian village. And so it goes on until, on day eight, you reach the village that, appropriately enough, is called Lofer.

What's green about this holiday? Cycling is the greenest way of touring around an area.

Who is it for? Just about anybody – you can always make it more sporting, if you want, by exploring further.

What's a green way of getting there? Hitch with Freewheelers (www.freewheelers.com – nothing at all to do with Freewheel Holidays) or go in your own car (via Eurotunnel) – if the car isn't full, use Freewheelers to offer the spare places; Eurolines coach from London to Salzburg; train from London to Salzburg. At Salzburg you'll be collected by mini-bus.

How much pollution am I saving? Compared with an air holiday to Austria, you could save up to half a tonne.

Anything not green about it? No.

What time of year? July and August.

What price? Expect to pay around £850 for seven nights B&B in three- and four-star hotels, including bike hire and luggage transport. On top of that you'll have to pay for lunch and dinner and the cost of getting there.

Where can I get further information/make a booking? Freewheel Holidays www.freewheelholidays.com
Tel. 01636 815 636.

Walking holidays

There's plenty of hiking in the UK to keep you occupied for years. But there's no reason you shouldn't get out and about on foot in mainland Europe, either. Especially if you live in the south-east, which means there's a whole lot of great Continental walking that's closer to you than Scotland.

Mainland Europe has some of the best walking and hiking in the entire world. Here's why:

- An extensive network of footpaths of all grades
- Good public transport to take you to your start point and collect you at the end
- Europe is safe
- It's easy to devise circular day-routes or long-distance itineraries that bring you to a delightful hotel every night
- There's wilderness and wilderness camping for those who want it.

Listing all the footpaths in Europe would take too much space but here are a few suggestions:

- **France** has a total of 40,000km (25,000 miles) of *Grande Randonnée* (GR) footpaths, most of which are in a good state of repair and well signposted. We'd recommend the GR223 coastal walk which you can begin at the ferry port of Cherbourg and continue, if you have the stamina, for 228km to Avranches. You could also take Eurostar to Paris and then set off along the GR22-22B to the splendid and famous Mont St Michel, an easy-going 288km. Slightly further away and more demanding is the GR10 traverse of the Pyrenees from Hendaye on the Atlantic coast to Banyuls-sur-Mer on the Mediterranean coast. Fit people with plenty of time do it straight off in about 40 days but others like to polish it off over three or four holidays, which gives a tremendous sense of achievement.
- **The Netherlands** has more countryside than you might imagine and you can enjoy the best of it along the *Gelrepad*, a hike in eight stages beginning at Enschede.
- **Belgium** has both coastal walking (the GR5A and the E9) and a surprisingly undulating route known as the *Transardennaise*, taking you 160km through the picturesque Ardennes.
- **Germany** has the wonderful Black Forest with an immense 23,000km network of well-maintained footpaths.
- **Spain** has done a lot to improve its footpaths in the past few years. The GR11 shadows the French GR10 along the southern face of the Pyrenees from Irún on the Atlantic coast to Cap de Creus on the Mediterranean coast. The official route guide allows 38 stages of a day each. We think the GR11 is more spectacular than the GR10 and the ideal would be to combine sections of each. Once you've reached Cap de Creus, the most easterly point in Spain, you could continue down the newish Costa Brava coastal path to Barcelona. Spain's most famous hiking route of all is the *Camino de Santiago*, the pilgrimage trail from Roncesvalles, near the French border, to Santiago de Compostela. If you do the whole thing you'll get a certificate at the end – and so you should because it's 730km.

If you're experienced walkers there's no reason you shouldn't make your own arrangements. That way you go at your own pace, stopping where you want to stop. But there's also a lot to be said for the camaraderie of a group, especially if you're alone (and you should certainly never be alone on the tricky stages of the GR10 and GR11).

What's green about this holiday? You can't get much greener than walking and, as you do it, you'll be helping to keep footpaths open.

Who is it for? There are so many different styles of walking you could say it's for just about everybody.

What's a green way of getting there? It's best to cross the Channel by tunnel and continue by train or coach. Don't fly.

How much pollution am I saving? By not flying to, let's say, Paris you'd be saving around 250kg of carbon dioxide for the return trip – plus what you save by hiking around rather than using motorised transport.

Anything not green about it? No.

What time of year? Summer in northern Europe; early autumn in the Alps and Pyrenees.

What price? A week's holiday could cost as little as £300.

Where can I get further information? Take a look at the National Tourist Office sites to begin with www.austria.info, www.belgiumtheplaceto.be, www.visitflanders.co.uk, www.franceguide.com, www.germany-tourism.co.uk, www.holland.com, www.visitportugal.com, www.tourspain.co.uk, www.myswitzerland.com; general tips on hiking at www.ramblers.org.uk.

How do I make a booking? Companies specialising in European hiking include:

→ www.inntravel.co.uk Tel. 01653 617 722
→ www.waymarkholidays.com Tel. 01753 516 477
→ www.foottrails.co.uk Tel. 01747 861 851
→ www.ramblersholidays.co.uk Tel. 01707 331 133
→ www.contours.co.uk Tel. 01768 480 451

WHY GO FURTHER NO. 14 – Dances with wolves?

For many people, wolves are synonymous with wilderness. Few sounds are as exciting as the howling of a pack at night. But you don't have to go to, say, Canada for the wolf chorus. You can hear them no further away than Portugal. What's more, if you follow our advice and take this trip we guarantee you'll not only hear the wolves but you'll see them, too. Very close.

You'll be in the rolling countryside of Estremadura, half-way between Lisbon and the coast, at the Iberian Wolf Recovery Centre. Here there are usually around 15 wolves, which have been either injured in the wild – often by cars or hunters – or rescued from captivity.

You'll stay in a rather dinky wooden chalet with other volunteers, if any. There's a room for four and another room for two (you can pay extra to have the small room to yourself if you're alone). This is one of the cheaper holidays of its kind, but you'll have to make your own way and buy and prepare your own food.

Your job could be monitoring the wolves and helping with feeding. You might also be asked to help with things like tree planting, general maintenance and, in summer, fire patrol. Weekends are free.

What's green about this holiday? It's vital work because this subspecies – rather smaller than the northern wolf – is on the brink of extinction.

Who is it for? You have to be 18 or over and willing to participate for a minimum of two weeks.

What's a green way of getting there? Hitch with Freewheelers (www.freewheelers.com); European Bike Express to Bayonne and then cycle the 625 miles (1,000km) from there (www.bike-express.co.uk Tel. 01430 422 111); car (via Eurotunnel) – if the car isn't full, contact Freewheelers (www.freewheelers.com); Eurolines coach from London to Lisbon; train from London via Paris. From Lisbon you can take the bus to Vale da Guarda (25km) where you'll be picked up.

How much pollution am I saving? If you'd flown to North America to see wolves you'd have generated an extra four tonnes of carbon dioxide or more.

Anything not green about it? This is completely green – as long as you don't fly.

What time of year? The Centre is looking for volunteers year round (winters are mild).

What price? This is one of the most inexpensive eco-warrior holidays, costing about £225 for the fortnight; on top you'll have to pay your own way there and buy and prepare your own food.

Where can I get further information? You can download a project file and read testimonials at www.ecovolunteer.org

How do I make a booking? There's an online booking form at www.ecovolunteer.org or you can book with Wildwings Tel. 0117 9658 333 www.wildwings.co.uk wildinfo@wildwings.co.uk.

HOLIDAYS WITH DOLPHINS

If we asked you to name something you'd like to do before you die, we'd bet that being with dolphins would come near the top of the list. There seems to be something very special about them. Maybe it's the mouth upturned in a perpetual smile. Maybe it's the extraordinary intelligence and the lure of interspecies communication. Or maybe it's the energy they seem to give off. Whatever it is, most people love them.

A trip boat is better than nothing but it's not the real way to see dolphins. Take our advice and you'll get closer than you ever imagined possible.

How would you like to ride on a small research vessel among the Croatian Islands, for example, getting to know the 160 or so bottlenose dolphins there? Every one of them can be identified by the researchers at the Losinj Marine Education Centre and has a name. Soon you'll be able to identify them as well. Your holiday would be to take down data (time, number of dolphins, activity and so on), maybe operate the hydrophone to record dolphin-speak, and enter up the data back at base. You'd also hear lectures on various aspects of dolphin behaviour from experts.

When you're not assisting, Losinj is a pretty attractive place with some of the clearest waters in the Adriatic.

What's green about this holiday? This is important work. Among other things, the data enables areas and fish species of particular importance to dolphins to be identified and, hopefully, protected.

Who is it for? Apart from a genuine enthusiasm for dolphins you'll need to be fairly fit, able to swim and over 18.

What's a green way of getting there? Hitch with Freewheelers (www.freewheelers.com) or go in your own car (via Eurotunnel) – if the car isn't full, use Freewheelers to offer the spare places; Eurolines coach from London to Venice or Trieste; train from London to Venice or Trieste. Once in the area there are ferries from Rijeka (the shortest hop), Venice, Trieste and Ancona (the longest). Buses will take you Trieste-Rijeka-Losinj or Zagreb-Losinj.

How much pollution am I saving? By not flying you'll be saving about half a tonne of carbon dioxide.

Anything not green about it? The vital work, and your money, more than make up for the energy used getting here – but no flying.

What time of year? Volunteers are needed from mid-June to the beginning of October.

What price? About £550 for 11 nights, including accommodation, food and lectures. You have to pay your own way there.

Where can I get further information? For the island itself see www.island-losinj.com.

How do I make a booking? Go to www.blue-world.org and download the booking form. For other similar holidays with dolphins take a look at www.fnec.gr/fiskardo.htm (based on Kefalonia, Greece) and www.delphismdc.org (for Italy).

Adopt a dolphin

Why not adopt a dolphin? Of course, you can't actually take it home with you. And, in reality, your dolphin won't benefit directly. But your money will go to help dolphin research and dolphin protection in general

If you do want to make things more personal, some organisations will name a dolphin after you or, indeed, anyone you choose. For example, the Bottlenose Dolphin Research Institute based in Sardinia will permanently enter the name of your choice into its database. Thereafter, while you're, say, sitting in your garden having a cup of tea you'll have the pleasure of knowing that somewhere off the coast of Sardinia a researcher is excitedly pointing at a shape in the water and shouting, "There's Hilda". Or Bert or whatever name you wanted. And when you join the research team for a few days, as we suggest [*see above*], you'll have the thrill of seeing and possibly even making eye contact with *your* dolphin. Adoptions with BDRI cost £40 a year at the time of writing and adoptions are £150 the first year and then £40 a year thereafter.

To learn more go to **www.thebdri.com/support/adopt.htm** and click on 'Dolphins available for adoption'. On the site you'll also find details of volunteer work (from around £35 up to £60 a day depending on length of stay and time of year).

A REAL BIRD IN THE HAND

Many of the Greek islands were discovered long ago and have now been completely spoiled. Others remain a secret. Have you heard of Antikythira, for example? Nor has anybody else. Antikythira lies about half-way between the southernmost peninsulas of the Peloponnese and the well-known holiday island of Crete. What's special about Antikythira is that in the spring and again in the autumn huge numbers of raptors migrate over it, onto it and from it. Sometimes you can lie on your back and see literally a score or more huge birds silhouetted against the blue sky. All the harriers can be seen here – marsh, hen, pallid and Montagu's – plus booted eagles, buzzards, honey-buzzards, black kites, red-footed falcons and Levant sparrowhawks. What's more, Bonelli's eagles, Eleonora's falcons and kestrels actually breed on the island.

That's not all that's special about Antikythira. This little island with a coastline of just 15 miles (24km) is a Natura 2000 site and protected under EU legislation. That should count for a lot although, sadly, it doesn't always. This is the Greek islands as they used to be. Just 50 or so inhabitants and very little development. If you want tranquillity, Antikythira is the place to come.

But we're not suggesting you just drop out and do nothing. Those birds need identifying and recording, an important task which compensates for the energy used getting here. If you can tell *Circus aeruginosus* from *Circus pygargus* you can stay in the Hellenic Ornithological Station (it used to be the school but there isn't much need for one now) and help out. If you can't, then get a field guide and study it. And if you're really interested in birds, the HOS also needs ringers. Each year volunteers handle around 4,000 birds of about 50 species. There are two places for beginners but if you actually have a licence you'll be even more welcome.

What's green about this holiday? Bird identification and ringing are important for conservation.

Who is it for? You need to be pretty keen on birds *and* solitude.

What's a green way of getting there? Hitch with Freewheelers (www.freewheelers.com) or go in your own car (via Eurotunnel and the Brindisi or Bari ferries) – if the car isn't full, use Freewheelers to offer the spare places; Eurolines coach from London to Athens, via the Brindisi/Bari ferries; train from London via the Brindisi/Bari ferries. From Athens (Piraeus) there's a ferry that takes nine hours or from Gythion (Yithio), the ancient port of Sparta, there's a ferry that takes three hours.

How much pollution am I saving? Travelling to Greece this way you'll save at least a tonne of carbon dioxide as compared with flying. In addition you'll be staying on an unspoiled island in very unspoiled (read simple) accommodation.

Anything not green about it? This is about as green as it gets – as long as you leave the flying to the birds.

What time of year? The HOS needs volunteers from around the end of March to around the end of May and again from mid-August through to the end of October.

What price? You'll have to pay your own fares and food; beyond that the cost is negligible.

Where can I get further information/make a booking? There are full details and an online booking form at www.ecovolunteer.org, or you can book with Wildwings Tel 0117 9658 333 www.wildwings.co.uk wildinfo@wildwings.co.uk.

GREEN CRUISING

Green cruising has nothing to do with your colour when the weather gets a bit rough. It has to do with a special style of cruising. Because, generally – as you'll have noticed in *Part One* – cruising isn't the least bit environmental. It seems as if it *must* be, given that sense of oneness with the oceans and nature generally. But, in energy terms, cruising is like taking a small town, complete with cinemas, theatres, swimming pools and restaurants, and moving it around. It's a disaster.

But that doesn't mean you can't *ever* enjoy, say, the Mediterranean from a moving deck. There is a way. *Sailing.* Now before you start objecting that you get seasick in small boats let us tell you there are *big* boats with sails. *Sea Cloud*, for example.

Sea Cloud is one of the most magnificent ships afloat. You may not have heard of her because she doesn't accommodate thousands of people. But she does accommodate 64 passengers in considerable style. So that's far from tiny. And – this is the important point – she *sails*. We're not talking about one of those new-fangled aerofoil things that cuts fuel consumption a few per cent on some otherwise ordinary cruise ships. We're talking about four-masts and real sails. The sort that Captain Bligh would have relished.

You yourself don't have to go aloft, we should stress. That's the work of the crew. All you have to do is enjoy the curve of the sail and the bow cutting through the water.

You're probably thinking that a real sailing ship is going to be extremely cramped and uncomfortable. Not a bit of it. Sea Cloud, you see, was built by the tycoon E.F. Hutton for his wife Marjorie Merriweather Post in the late 1920s. And her guests wouldn't have liked sleeping in hammocks. No, Marjorie personally undertook the interior design, fitting out the interior with oak panelling, marble bathrooms, parquet floors and valuable antiques.

So what more could you want? As to where could you go, you're going to have to forget the Caribbean. In order to keep things green you can't fly out. Instead, take the train to Nice, board *Sea Cloud* and set off for Italy, Elba, Sardinia and Corsica.

What's green about this holiday? A lot of the time you'll be sailing, not motoring.

Who is it for? This is more for adults than children and, of course, you have to be able to afford it.

What's a green way of getting there? *Sea Cloud* sails various itineraries with different departure and arrival ports, served either by rail and/or coach.

How much pollution am I saving? Compared with a conventional cruise ship you could easily save 1.5 tonnes of carbon dioxide in a week.

Anything not green about it? They run the engines some of the time.

What time of year? Summer.

What price? A week aboard *Sea Cloud*, full-board including wine, beer and soft drinks, will cost from around £2,300 per person.

Where can I get further information? You can read all about the ships at **www.seacloud.com**

How do I make a booking? **www.noble-caledonia.co.uk**
Tel. **020 7752 0000**

*

As we've said, cruise ships just aren't green. They use far too much energy. But if you insist on a conventional cruise liner at least you'll be a little bit greener on a Windstar ship. A little small by today's supercruiser standards, the Windstar ships nevertheless look fairly conventional except for one thing. They have masts. Proper ones. Sixty metres high. Four of them. And on them they carry up to six self-furling sails amounting to some 21,500 square feet of Dacron. That can save a lot of fuel. And look very nice, too.

What's green about this holiday? The ship is sail-assisted.

Who is it for? Apart from the sails, this is pretty much like a regular cruise ship.

What's a green way of getting there? You can get to the Mediterranean and Baltic ports by train and/or coach.

How much pollution am I saving? It all depends on the route and the wind but probably around half a tonne of carbon dioxide in a week.

Anything not green about it? The engines still get quite a lot of use.

What time of year? Spring to autumn.

What price? From about £1,250 for seven days.

Where can I get further information? You can read all about it at www.windstarcruises.com or e-mail info@windstarcruises.com.

How do I make a booking? At the above website or by telephoning Seattle, USA 00 1 866 766 3873.

*

Every autumn herrings shoal in vast numbers in the fjords of northern Norway, creating one of the great wildlife spectacles in the world. Literally hundreds of orcas (killer whales) arrive to feed upon them and there are white-tailed sea eagles and otters as well. Throw in the Aurora Borealis (Northern Lights) and you have everything you could hope for. Best of all, you'll be watching from a two-masted schooner called the *Noorderlicht* and, wind permitting, creating very little pollution.

The *Noorderlicht* is a reassuring 46 metres long and was comfortably refitted in 1991. The deckhouse contains a cosy settee for wildlife watching when the weather is bad. Below is a spacious saloon and 10 cabins, each with two single berths and a basin. And here's another great thing. You actually get to help sail the boat and take a turn at the wheel (although there are three professional sailors to do all the tricky stuff).

What's green about this holiday? You'll be sailing (wind permitting); you'll also be demonstrating that whales have value alive.

Who is it for? You'll need to be a little bit adventurous.

What's a green way of getting there? The greenest way is by train and bus but if you live in the north of England the ferry from Newcastle to

Stavanger/Bergen (then onwards by train/bus) would be acceptable. The boat is based at Lodingen. For a couple, ferry prices start at £112 and there's an automatic Norwegian Environmental Charge of £1 per person on top.

How much pollution am I saving? Compared with flying to Norway and then cruising aboard a conventional liner you might save around one tonne of carbon dioxide.

Anything not green about it? No.

What time of year? Winter (October to December) if you want to see orcas.

What price? £770 each, based on two sharing a cabin, including all meals and shore excursions, but not including travel to the port of Lodingen.

Where can I get further information/make a booking?
www.wildwings.co.uk Tel. 0117 9658 333

*

This next holiday combines a comfortable sailing boat *and* cycling, all in one holiday. So it's sort of doubly green. The boat is a *bahriyeli* – think of a Turkish *gulet* only bigger (three masts) and more luxurious (eight to 10 cabins with air conditioning, shower and toilet). The idea is to cruise the islands of the Greek Aegean under sail (assuming there's enough wind) and to explore their interiors on two wheels. You'll sail from Marmaris, Turkey, and make your first landfall on Symi, about five hours away. It's an enthralling introduction to the islands. The entrance to the natural town harbour of Gialos is flanked by a riotous jumble of houses, painted in gaudy lemons and peaches, like Byzantine versions of Bath. Here you'll spend a day-and-a-half exploring before cruising on to the volcanic island of Nisiros. Finally, via Tilos, you'll come to the famous holiday island of Rhodes but you'll avoid the crowds most of the time, as you pedal from the idyllic anchorage at Kamiros Skala across the middle of the island to the equally superb Lindos. It's the longest punt of the trip at 50 miles, but well worth the effort.

What's green about this holiday? It combines cruising under sail and cycling.

Who is it for? You need to be moderately fit.

What's a green way of getting there? The train is the best option. Since it takes three nights you should plan to make it part of the holiday. Alternatively, you could hitch with Freewheelers (www.freewheelers.com) or go in your own car (via Eurotunnel) – if the car isn't full, use Freewheelers to offer the spare places. It's about 1,900 miles (3,000km) each way. You could cut the journey by taking one of the ferry services from Italy but most take so long you'll end up creating more pollution than by flying – the Brindisi/Çesme (near Izmir) service is just about acceptable.

How much pollution am I saving? Compared with a holiday to Turkey by air you could save about one tonne of carbon dioxide.

Anything not green about it? Given the distance, you're inevitably going to create some pollution.

What time of year? Summer.

What price? Around £750 for seven nights, not including getting to Marmaris.

Where can I get further information? Turkish Culture & Tourism Office, 4th Floor, 29-30 St James's Street, London SW1A 1HB Tel. 0207 839 7778; or see www.gototurkey.co.uk or e-mail info@gototurkey.co.uk

How do I make a booking? http://2wheeltreks.co.uk
Tel. 0845 612 6106

GREEN SKIING

Some people are passionate about skiing and snowboarding. We are. So like you we'd be pretty reluctant to give it up in the name of the environment. The problem is that skiing concentrates large numbers of people in delicate environments, can use scarce water resources for artificial snowmaking, disturbs wildlife at the most vulnerable time of year, uses a lot of energy, disfigures the wilderness with ski lifts and ski runs and, quite often, increases the risk of avalanches because of tree clearing.

It's quite a list. The irony is that we who love winter sports are partly responsible for damaging the very mountain environment we've come to see. So there has to be a rethink. No more development, no more ski lifts, no more ski runs.

There is a solution. And that's to do it the old-fashioned way. Without lifts. And without prepared runs. At first it may sound too much like hard work for too little reward. But just stop for a moment and analyse what it is you like about being out in the mountains. Sure you enjoy the thrill of carving those turns down the slopes. But don't you also enjoy just *being* there. Of course you do. The pristine snow, the trees, the winter scents, the physical exertion and, if you're lucky (and you usually are) the clear blue skies. Okay, you won't get in so many vertical feet but you'll have all the rest. Plus adventure and tranquillity.

Nowadays there are five main styles of getting away from the resorts:

- **Cross-country skiing**. You could see this as just a speedy way of getting along in the snow. You don't tackle steep slopes at all. Instead, you seek rolling or even flat terrain. It's easier than downhill skiing, although it's not without its skills, and you won't get frightened. The skis and boots are pretty light.
- **Telemark skiing.** Enthusiasts think this is the ultimate. Not everyone agrees. Essentially, telemark equipment is lighter than downhill (normal skiing) equipment but more sturdy than cross-country equipment. The name comes from a special way of turning (telemarking), in which you have to adopt the position of a fencer lunging at someone. It's pretty hard on the knees at first. The advantage over cross-country equipment is that you can tackle the difficult slopes.
- **Mountain touring skis.** These look pretty much like ordinary downhill skis but the binding is rather different, allowing you to lift your heel up when you want. With ordinary downhill skis you can't do that. If you want to know why, just try walking up a hill whilst keeping your heels on the ground. Exactly! But for the descent you lock your heels down and you then have far more control than with either telemark or cross-country skis.
- **Snowboards.** If you've not yet made the acquaintance of snowboards we're talking about something that resembles a surfboard. The

problem with snowboards is that, unlike skis, you can't walk around whilst wearing them. Which means you'd have to hike up the mountain in your boots (not easy if the snow is soft), carrying the board on your back. Now that *is* hard work. And very, very few people ever did it. But some clever engineers have come up with a solution. Split boards! Essentially the board is in two parts. For the ascent it's like walking on skis. But when you get to the top you fix the two parts together and you then have a snowboard on which you can swoop down.

- **Snowshoes.** You can't really slide very much on snowshoes. But if you want to get out in the snow and don't fancy any of the above then these could be for you. If you can walk you can snowshoe.

So, really, you haven't got any excuse to use those lifts.

But – you may object – isn't it dangerous to go skiing away from the pistes? Well, in the first place, avoiding downhill resorts doesn't necessarily mean avoiding established routes. Most cross-country skiers and snowshoers follow trails that have been specially prepared and which are known to be safe. As regards skiing and snowboarding away from marked trails, this is something that has to be put into perspective. Of course there are places it's dangerous to go off-piste. But only in the same way that there are places it's dangerous to hike in summer. Just don't go anywhere near those places. Stick to safe conditions and routes that are known to be safe in those conditions. Safety is a big subject and we can't deal with it all here. So get a specialist book, take lessons and *make your initial trips with an expert*. Then you can retrace those same routes on your own, if you want – always provided the conditions are at least as good, or better.

Here are the basic rules:

1 Keep away from avalanche slopes. Even experts find it hard to predict when a slope will avalanche but there are some clear guiding principles. First, don't go when the meteorologists have announced a high avalanche risk generally. Second, always wait a few days after recent snow so it has time to settle down and consolidate. Third, don't go if the snow is unusually deep – it's much harder for shallow snow

cover to avalanche. Fourth, never go on a slope that has *ever* avalanched in the past (if you're not familiar with the area, seek local advice).

2 There should always be at least three of you – one to stay with an injured person and the other to go for help.

3 Start with easy things, close to 'civilisation' and hone your technique before moving on to anything more demanding.

4 If there's the slightest danger, go back. Spread out until you're in a safe place again – that way, if one of the party gets into difficulties the others, who are unaffected, will be able to take appropriate action.

What's green about this holiday? Skiing/snowboarding without using lifts or prepared runs is much better for the environment.

Who is it for? If you're reasonably fit, then one of the various styles – from snowshoeing through to ski mountaineering – will suit.

What's a green way of getting there? As always, the rule is to stick to land transport – train, coach or car with all seats taken.

How much pollution am I saving? Compared with flying to the Alps you could save about 750kg of carbon dioxide. By keeping away from ski stations you won't be contributing to tree felling to make runs, or to unsightly ski lifts.

Anything not green about it? Even without using ski lifts you still need to take care; the mountain environment is fragile.

What time of year? It's easier later in the season when the snow has had plenty of time to consolidate. If it's warm, ski early in the day and be down before the snow becomes too wet and soft.

What price? Ski lift passes tend to cost around £100-£150 a week, so you can save quite a bit this way. Otherwise, it's the same as for a normal ski holiday. Waymark is offering eight days in Seefeld (excellent for cross/country skiing) from £515; Headwater is offering the Jura from around £700 self-drive; the Telemark Ski Co is offering the Oetztal High Level Route (experts only) at £830 for one week. Note that you can often get last-minute ski deals from as little as £300 a week self-catering.

Where can I get further information/make a booking? These are some of the specialist companies:

- → www.headwater.com Tel. 01606 720099
- → www.hfholidays.co.uk Tel. 020 8905 9556
- → www.telemarkskico.com Tel. 01248 810337
- → www.waymarkholidays.com Tel. 0870 9509800

WHY GO FURTHER NO. 15 – Chimpanzees

Chimpanzees are among the animals people go on safari to see. But it certainly isn't necessary to go to Africa. We're suggesting you should go no further than northern Spain. There, near to Girona, is a chimpanzee rescue centre that's always looking for volunteers.

But, you may object, that's hardly the same thing as seeing them in the wild. To which we can only say, no it isn't. It's *better*. Okay, not in every way. But just consider this. If you volunteer you'll get closer to chimps than you ever could in the bush. You'll be able to observe them at close quarters and even prepare and distribute their food.

The Chimpanzee Rescue Centre takes both long-term volunteers and holiday volunteers. That's where you come in. If you've got just a couple of days to dedicate to this then you're in for the experience of a lifetime. You can easily combine those two days with a beach holiday on the Costa Brava or an adventure holiday in the Pyrenees. And, if you do, you'll benefit the chimpanzees directly as well as help the centre to continue its good work.

What's green about this holiday? You'll be helping chimpanzees without going any further than northern Spain.

Who is it for? You need to be the kind of person who's willing to muck in – and sometimes muck out, too.

What's a green way of getting there? Eurolines coach or train to to Girona; hitch with Freewheelers (www.freewheelers.com) or go in your own car (via Eurotunnel) – if the car isn't full, use Freewheelers to offer the spare places; European Bike Express to Ampuriabrava and then cycle the 50-odd miles (80 km) from there (www.bike-express.co.uk Tel; 01430 422 111).

How much pollution am I saving? Compared with flying to Girona, you could save around 500kg of carbon dioxide; compared with flying to central west Africa you could save three tonnes or more.

Anything not green about it? No.

What time of year? Volunteers are needed all year.

What price? You'll pay £160 for two days (or £270 for two people) with your travel and accommodation on top.

Where can I get further information? Take a look at www.fundacionmona.org and www.mona-uk.org or telephone the UK representative on 01223 210 952.

How do I make a booking? You can also make arrangements through www.responsibletravel.com

Chapter 11

Green holidays in the rest of the world

Basically, long-distance holidays aren't green unless you get there by land transport. If you fly or cruise you can't call yourself green unless you've got some extremely compelling, environmentally beneficial reason to go. And only then if there's no alternative. And, even so, you've still got to arrange copious offsets.

So, as you can guess, we're not describing many long-haul holidays here. The ones we do describe are mostly by land or – wait for it – sailing ship. We're only going to allow flying when there's a very special reason.

And when you stop to think about it there aren't many good reasons to travel long-distance. Yes, the idea of shopping in New York may be appealing but you can't buy anything in New York that you can't also get in London. (If you think you do know of something, just have them send it.) And, yes, it's fascinating and wonderful to see Machu Pichu but have you yet been to Stonehenge? Or puzzled over the paintings in the caves at Niaux in the French Ariège? Or watched whales in Scotland? In reality, there are plenty enough man-made and natural wonders close to home to keep you occupied for a lifetime.

Not that every holiday has to be mentally invigorating, spiritually uplifting or environmentally beneficial. After a hard year you're perfectly entitled to chill out (or is it heat out?) on the beach. Just don't do it in the Caribbean.

So are the days of long-distance holidays over? No. You just need a different approach. The Alexandra-based Greek poet Constantine Cavafy had something to say on the subject:

When you set out on the voyage to Ithaca
Pray that your journey may be long
Full of adventures, full of knowledge.

In other words, if it's not actually better to travel than to arrive it's certainly of equal importance. Make the journey part of the holiday. Then if it takes a long time it doesn't really matter. On the contrary. The longer the journey, the more pleasure you'll get from it. If you haven't got the time right now, postpone the trip until you do. In the meantime, stick to destinations that are close by.

Just send the money

Before signing up for a long-distance 'environmental' holiday, give very careful thought to your motives and its impact. The good that you're doing will quite likely be more than undone by the energy you're using to get there. And, in many cases, there are local people who can do the jobs just as well, if not better, than those who would have to fly half-way round the world to get there.

What these organisations really want and need is *money*. So before you sign up for a £1,500 volunteering holiday in, say, South America, ask yourself this. Would it be better if I stayed in Europe and volunteered for some important work nearer to home? The sums could work out like this:

- Amount saved by holidaying in Europe – £750
- Pollution saved by not flying to South America – 8 tonnes carbon dioxide

You could then send, say, £200 to the original project, still be better off yourself, still do something worthwhile on holiday *and* protect the planet from a whole lot of pollution. That way, everybody wins.

TAKE YOUR TIME – TAKE A SABBATICAL

Anyone going a long way normally thinks in terms of travelling fast. But that's not a green approach. If you want to visit, say, Australia then you have to go *slowly*. And why not? It's only recently that human beings have had the ability to take a two-week vacation Down Under. Maybe in the not too distant future we'll all have that chance again. But right now it's unsustainable.

As we already saw in *Part One* it *is* possible to get to the furthest corners of the globe. It just calls for a different mental approach. The idea of taking, say, a year's break from work is gaining ground. *That's* the time to enjoy long-distance travel. Or during a gap year. Or after retirement. Then you can make your way by coach, train, expedition vehicle, sailing ship, bike or even by hitch-hiking. A two-week holiday is very definitely not the time. For those occasions, stick to something nearer home.

GREEN OVERLAND HOLIDAYS

Remember how it was years ago? Everyone used to pile into a van and set off for South Africa. Well, those days are back. In fact, despite cheap air fares, they never really ended. But what's happened recently is that, given the environmental situation, overland treks have once again become the thing to do. You can *still* go on safari, you can *still* see gorillas in the mist, you can *still* visit the Pyramids. But, for environmental as well as practical reasons, you do it by land.

The typical expedition vehicle

Different companies equip their vehicles in different ways. There's no mass production line churning these things out. So before you commit to an overland trip it's a good idea to get full details of the vehicle. After all, if you're doing one of the longer trips you'll be spending the equivalent of three months inside it.

Oasis Overland, for example, mostly uses Scania 93 series 4 x 2 trucks of between 220 and 280 horsepower and they're a good benchmark. On Africa's tracks – and, indeed, off-track – they're the kind of get-you-out-and-get-you-home tank you can rely on. Oasis's vehicles have forward and inward facing seats so you can be convivial when you feel like it. Of course, they've been converted and large windows installed so everyone has a good view. And as a special feature they have what's known as 'The Beach' – basically an exterior area. Look out, too, for features such as 12V charging points so you can keep your mobile topped up. Naturally, the trucks are equipped with chains and sand mats, for when you get stuck in the sand – as you inevitably will.

Typically, African overland treks kick off in Spain, so you'll have to get yourself there. By land, of course. Each company has its own itinerary but broadly they all follow a similar circuit. Which is across to Morocco, down the west side of the continent to Cape Town and back up the east side, finishing somewhere like Istanbul or Cairo.

We're talking something like 26,000 miles (42,000km) and 30-odd countries. And if you do go the full distance you're going to have to devote something like 10 months of your life to the project. So, in that well-known phrase, it's the trip of a lifetime. But on this occasion it happens to be true.

No air tour could possibly be as exciting or stimulating or give you such an intimate understanding of Africa. You'll see things 'ordinary' tourists can only imagine. You'll be in the middle of the action. The Riff Mountains, the Sahara, the Niger River, the Nile, mountain gorillas, lions, cheetahs, elephants, hippos, the Masai Mara, Etosha – it's a list of at least 50 of the top 100 things you must do before you die. And all on one trip.

Okay, it's not luxury. A lot of the time you'll be camping far from any town. For days on end you'll be out of communication with the wider world. But it's the real deal. It's certainly not just for young people. (In fact, older people, with their healthier bank balances, tend to predominate). And it is the most environmental way of going.

What's green about this holiday? You're seeing all those long-distance things you dreamed of but *by land.*

Who is it for? You need to be someone with plenty of time and an ability to rough it.

What's a green way of getting there? You can get yourself to the start point in Andalucia (Malaga or wherever) by hitching with Freewheelers (www.freewheelers.com) or by Eurolines coach or by train.

How much pollution am I saving? Compared with flying to and from South Africa and points in between you'll be saving five tonnes or more of carbon dioxide.

Anything not green about it? The distance might seem to be unenvironmental but when you divide the total by the number of passengers it's almost certain to work out well under the mileage you'd do in the UK in the same time.

What time of year? These trips take most of the year.

What price? For the full trip think in terms of £3,500 for the package plus £1,500 for obligatory expenses on the ground (overland companies tend to charge them separately) plus personal spending money. It seems a lot at first – but not when you compare it with 10 months or so living in Britain.

Where can I get further information? Put 'overland trek' into your search engine.

How do I make a booking? Two companies worth looking at are www.footprint-adventures.co.uk Tel. 01522 804 929 and www.oasisoverland.co.uk Tel. 01963 363 400.

WORTH THE POLLUTION?

If nobody goes to see wild places and wild animals will they still exist? We argued in *Part One* that they won't. In an ideal world we'd all, rich and poor alike, be willing to make a financial sacrifice for the poetic benefit of knowing that wilderness and wild animals are out there, even if we can't see them. But we don't live in an ideal world.

So we have to support conservation efforts in a different way. By spending money.

Take Costa Rica, for example. Since 1963 the country has created 70 protected areas covering over one-fifth of its territory. And they don't just exist in name. Logging has been stopped in these areas and, in some cases, local people have been relocated. This wouldn't have been possible without the income that ecotourism can generate.

That's not to say ecotourism doesn't create problems of its own. The number of tourists visiting Costa Rica has been going up at 6 per cent a year and the pressure on some of the most popular parks has damaged plants and turned some animals, especially monkeys, into scavengers of human rubbish.

That's a pity, but would a mine be better? Would logging be better? There's no need to answer that. Of course, *nothing* would be best of all. Just leave the wilderness as wilderness and shut out all human activity. But, with the exception of a few very special and delicate areas, that just isn't going to happen.

Ecotourism, if it's genuine, is the least worst.

Here then are some long-distance holiday ideas that we think are worth the pollution. As always, of course, the less energy that goes into getting you there the better. We begin with some of the more conventional and comfortable kinds of holidays on which your contribution to the environment is mostly financial. And we finish with some of the more unusual ones where your contribution will be practical, too – tracking, recording data, rescuing, feeding, cleaning ... getting exhausted. But it's worth it.

HOLIDAYS WITH THE ENVIRONMENTAL INVESTIGATION AGENCY

The Environmental Investigation Agency (EIA) is a non-profit making conservation body that works to save endangered species. In Indonesia, for example, the EIA has been working together with local conservationists to try to save the orang-utan. And that's the first of our long-distance air holidays.

BORNEO'S ORANG-UTANS

The world population of orang-utans has fallen by half in just 10 years, a situation that has to be reversed urgently. If you go on this holiday you'll play your part because 10 per cent of your money will go to the EIA to help it continue its work. Another chunk will go to the local people in Borneo who provide services, food and accommodation, thus clearly demonstrating the financial benefits of conservation. The highlight of the holiday for many will be the Sepilok Orang-utan Rehabilitation Centre which cares for orphaned orang-utans and prepares them for release back into the wild. You'll also stay at the Sukau Rainforest Lodge where you should see orang-utans in the wild, as well as proboscis monkeys, hornbills, kingfishers, Asian elephants and the swiftlets that nest in the nearby caves. Finally, you'll go to the Danum Valley Conservation Area, probably the best natural history site in all of south-east Asia.

Does it really achieve anything?

Lots of organisations ask for money and publish pretty brochures but does anything really ever get done? In this case, yes. In fact, there are often laws protecting habitats and wildlife but they get ignored. This is where the Environmental Investigation Agency (EIA) comes in, working undercover to gather evidence and then presenting it to the authorities.

In 2005, for example, the EIA, together with local conservation organisation Telapak, produced a report into illegal logging which led to a government crackdown. More than 400,000 cubic metres of stolen logs and sawn timber were seized, along with logging equipment, trucks and ships. More than 170 people were arrested, including members of the armed forces and police.

The forests are still not safe from logging companies but it was an important step. You'll be supporting the EIA by going on this holiday. It would be even better if you could just send a donation, instead. But if you want to see those orang-utans, the holiday is a reasonable compromise. Other EIA tours include the Panna Tiger Reserve in Madhya Pradesh and the Kaziranga National Park in Assam.

What's green about this holiday? You'll be helping to save orang-utans from extinction.

Who is it for? Everyone interested in wild places and wild animals.

What's a green way of getting there? Realistically, you're going to be flying.

How much pollution am I saving? You'll actually be creating a massive seven tonnes or more of carbon dioxide on your return flight but if it helps reduce the destruction of the rainforest it could be worth it.

Anything not green about it? The flight.

What time of year? Usually September.

What price? £2,295 for 11 days.

Where can I get further information? You can read all about the EIA at www.eia-international.org

How do I make a booking? www.naturetrek.co.uk
Tel. 01962 733 051.

HOLIDAYS WITH THE ZOOLOGICAL SOCIETY OF LONDON

The Zoological Society of London (ZSL) has a rather quaint air to its name. But it means business. Founded almost two centuries ago by Sir Stamford Raffles, the ZSL is nevertheless at the cutting edge of nature protection.

TANZANIA'S CHEETAHS

Cheetahs, like dolphins, are emblematic animals. Ones whose loss, even if you've never seen them in the wild, would deeply move you. If you take this 12-day tour you'll not only see the cheetahs of the Serengeti you'll actually be helping the ZSL to protect them.

You'll visit Arusha, one of Africa's most beautiful national parks, spectacularly set between the volcanic peaks of Mount Meru and the iconic Kilimanjaro. In the rainforest around the Ngurdoto Crater you'll see monkeys and in the Momella Lakes hippos and thousands of flamingos. Next you'll drive to Tarangire National Park where, in the rolling bushland dotted with baobab trees, you'll see elephants, buffalos, eland, oryx and impala. Later, at the legendary Ngorongoro Crater you'll observe zebras, jackals, spotted hyenas and lions. Finally, you'll reach the grasslands of the Serengeti where you'll have your best chance of finding cheetahs. Altogether it adds up to probably the greatest wildlife spectacle on earth.

Does it really achieve anything?

Over the years the ZSL has reintroduced the Arabian Oryx into the wild, achieved the world's largest reintroduction of captive-bred mammals (sand gazelles in Saudi Arabia), carried out the first radio-collar study of the Sumatran tiger and helped establish 15 Marine Protected Areas for seahorses in the Philippines. These, and many other projects, result in real, tangible improvements in the environment. Ten per cent of the cost of your holiday goes to the ZSL and, at the same time, the money you pay locally provides the justification for preserving these magnificent wilderness areas.

What's green about this holiday? You'll be helping fund the work of the ZSL and also giving value to wildlife and wilderness.

Who is it for? Anyone who loves wildlife – and you won't have to rough it.

What's a green way of getting there? Realistically, you'll be flying.

How much pollution am I saving? If you fly, you'll be responsible for about seven tonnes of carbon dioxide – but cheetahs need your support.

Anything not green about it? The flight.

What time of year? The British winter.

What price? You'll be paying around £3,000 for 12 days.

Where can I get further information. You can read all about the marvellous work of the ZSL at www.zsl.org

How do I make a booking? www.naturetrek.co.uk Tel. 01962 733 051

HOLIDAYS WITH INTERNATIONAL ANIMAL RESCUE

International Animal Rescue is a Sussex-based charity dedicated to the rescue and rehabilitation of suffering animals all over the world. In particular, the IAR rescues sloth bears which are made to dance on the streets of India.

AGRA'S SLOTH BEARS

You'll spend a whole day at the Agra bear sanctuary, get behind the scenes and see how sloth bears that had been broken down are given back their health, energy and zest for life. It's heartening to see these endearingly shaggy, man-sized animals restored and you'll come away inspired. (By the way, sloth bears are not sloths – the name comes from an early 19th century mistake in categorisation due to their long sloth-like claws.) Your trip will also include visits to the Bharatpur Bird Sanctuary and the Ranthambore National Park as well as the Taj Mahal.

Does it really achieve anything?

The IAR has rescued nearly 250 bears since 2003 and rehabilitated them in three private sanctuaries. Its highly pragmatic approach also includes retraining the bears' former owners in new trades or professions so they won't simply capture other bears and begin again. It's estimated that 600 bears remain to be rescued.

What's green about this holiday? Ten per cent of the cost of your holiday will be helping to fund the work of International Animal Rescue and your local spending will give value to the conservation of wildlife.

Who is it for? Anyone who loves animals.

What's a green way of getting there? Realistically, you'll be flying.

How much pollution am I saving? By flying to India you'll be generating around five tonnes of carbon dioxide; on the other hand, your money is vital to animal protection and conservation.

Anything not green about it? The flight.

What time of year? The British winter.

What price? Around £1,550 for 10 days.

Where can I get further information? You can learn all about the IAR at www.iar.org.uk

How do I make a booking? www.naturetrek.co.uk Tel. 01962 733 051

WHY GO FURTHER NO. 16 – Bear essentials?

This is the first of our less-comfortable, more unconventional holidays. Ones in which you get up close and personal with animals. In this case it's bears (and there's a chance of seeing wolves, too).

When we think bear we normally think grizzlies in North America or polar bears in the Arctic (or perhaps, as described above, sloth bears in India). But you don't have to go anything like that far. Aside from the bears at the London Stock Exchange, the nearest wild bears are in the Pyrenees, but you have zero chance of seeing them. There are more in Eastern Europe, but the likelihood of a sighting is still slim.

In this holiday we're not offering you the *chance* to see bears. We're offering you the *certainty*. Without going outside Europe. And what's more, we're offering the opportunity to handle and feed bears. Because these are cubs that have been rescued (often after hunters have killed their mothers) and which will be released back into the wild at a year old. By that time, of course, they'll be pretty big and strong. So you'll have to be fairly tough yourself.

This work goes on at a biological station about 280 miles (450km) north-west of Moscow in the beautiful and wild Valday Hills. It's what they call taiga – virgin spruce forest together with mixed broadleaf, peat bogs, swamps, lakes, rivers and meadows. Wildlife is abundant and includes wolves, lynx, mink, beavers, elk, cranes and black storks.

The other work of the biological station is radio tracking. For that you'll need to be fit enough to hike long distances but, unlike caring for cubs, it can never be guaranteed that you'll actually see the bears (nor, in some cases, wolves) you're following.

What's green about this holiday? You'll be helping to save bears.

Who is it for? You need to be a no-nonsense country type over 18. The project staff recommend you should be under 45 but if you're a fit older person there's no problem.

What's a green way of getting there? Although it's quite a long way, it's perfectly possible to make the journey by coach or train. The most direct route is to Riga in Latvia from where you can take the train to Velikie Luki, inside Russia to the south-west of the Valday Hills. From there a car will take you to the research station. You can also reach the Valday Hills from St Petersburg or Moscow.

How much pollution am I saving? If you travel overland to the Valday Hills you could save three tonnes or more of carbon dioxide as compared with flying to Canada to see bears.

Anything not green about it? No.

What time of year? February to April if you want to help raise the cubs; May to October for tracking.

What price? About £900 for the minimum period of one month, plus transport costs.

Where can I get further information? www.ecovolunteer.org

How do I make a booking? There's an online booking form on the ecovolunteer website or you can arrange everything through UK operator Wildwings www.wildwings.co.uk Tel. 0117 9658 333

DANCES WITH ELEPHANTS

On this holiday you'll be doing things you never would have believed possible. How about this, for example? Would you ever have believed you could be friends with an elephant, accompany it to a lake for a swim and there scrub it? No, and nor will your friends believe you when you get back. So have your camera ready as proof.

The elephants, no longer needed for logging and rescued from a life of entertaining people on the streets, live at a sanctuary about 160km southwest of Bangkok near the Kaeng Krachan National Park. There you'll help take care of the elephants, feeding them, washing them, accompanying them on foraging expeditions and taking them for their swim. It's an important project and all the more so considering that only around 2,000 elephants survive in the wild in Thailand.

You'll have to make your own way to the sanctuary. If you're already going to Thailand, consider this as a way of making up for some of your transport emissions. At the sanctuary you'll have a simple room but usually with shower and toilet.

What's green about this holiday? Your practical help and financial contribution are important for the ongoing work of the elephant sanctuary.

Who is it for? Anyone who wants to get close to elephants.

What's a green way of getting there? Realistically, you'll be flying.

How much pollution am I saving? By flying to Thailand you'll be adding over seven tonnes of carbon dioxide to the atmosphere so do your best for the elephants when you get there.

Anything not green about it? The flight.

What time of year? Any time.

What price? Your stay at the sanctuary will cost you around £250 for the first week and about £175 for an extra week, including lunch and dinner.

Where can I get further information? www.ecovolunteer.org

How do I make a booking? There's an online booking form on the ecovolunteer website or you can arrange everything through UK operator Wildwings www.wildwings.co.uk Tel. 0117 9658 333

CAMEROONED

One of the greatest threats to habitat and wildlife comes from the conflict with local people. In wealthy Britain we apparently can't even tolerate the financial damage allegedly caused by badgers, so in Cameroon it's pretty understandable that local people sometimes find it hard to get along with elephants, leopards and lions. That's where you come in. By taking part in the work of the Benoue National Park in Cameroon's North Province you'll also be giving local people a reason to put up with elephants trampling through their crops. In a word, money. What you spend gives that wildlife tangible value to those people. And that's vital. Because otherwise those animals will disappear.

And what wildlife! This is a UNESCO World Biosphere Reserve and during your stay you're likely to see not only elephants, leopards and lions but also buffalo, antelope, warthogs, monkeys and masses of birds. You'll stay in a beautiful, traditional buckaroo with full board. So this is the real thing. And your task will be to support the research in various ways, such as making observations or recording data on a computer.

What's green about this holiday? You'll be helping the survival of wildlife in Cameroon.

Who is it for? This is one of the few such projects on which children are allowed but it certainly isn't recommended under the age of six. You'll all have to be fairly tough.

What's a green way of getting there? You could go with one of the overland trekking outfits but, given the importance of the project, flying is permissible.

How much pollution am I saving? If you do take the overland option you'll be saving at least three tonnes of carbon dioxide compared with flying. If you do fly just make sure you do a good job when you get there.

Anything not green about it? Flying there.

What time of year? Any time.

What price? About £1,000 for the minimum of three weeks, including full board.

Where can I get further information? www.ecovolunteer.org

How do I make a booking? There's an online booking form on the ecovolunteer website or you can arrange everything through UK operator Wildwings www.wildwings.co.uk Tel. 0117 9658 333

Sailing round the world

How can you get to places like the Caribbean, or South America or Australia, or go round the world, in a green way? That's to say, without either flying or going on a cruise ship? Impossible? Not at all.

The solution is a sailing ship. Yes, they're the past but they're also the future. We've already looked at sailing ships in Europe. And there are quite a few. Sailing ships that go round the world are a bit harder to come by. But they exist. And if people have any sense there'll be a lot more soon.

The Picton Castle is one of them. She (sailing ships are always 'she') doesn't follow the same itinerary every year. Sometimes she goes round the world. Sometimes she goes round something smaller. But she usually stops in Britain so you can get aboard. Around the time this book was published, for example, she was off on a year's voyage of the Atlantic, starting in Nova Scotia but also taking in Britain before heading off for Scandinavia, France, Spain, Morocco, Madeira, the Canary Islands, the famous crossing of the equator, Brazil and the Caribbean. For most people, these last two will be the highlight, including the 21-island archipelago of Fernando de Noronha, Grenada, Anguilla, Bequia, Cariacou, Dominica, Guadeloupe, Petite Martinique, Nevis, Barbados and the British Virgin Islands.

What's life like aboard? The first thing to understand is that you're not a passenger, joining in when you feel like it and sunbathing when you don't. You'll be a member of the crew, following a watch system of four hours on and eight hours off. In reality, that's not hard. It amounts to an eight-hour day, giving you plenty of time for relaxation and rest – although you'll have to get used to some unusual sleep patterns. And you'll be backed up by a professional crew of 12. But, in any event, taking part in the running of the ship is a privilege. And they even let you steer. Great fun!

Nor will you have a private cabin. A private space, yes. That's to say, a generous bunk with privacy curtains and your own light. Some stowage space. And, well, that's about it.

But there's something about shipboard routine – on a real sailing ship, that is – that makes it all seem very natural.

What's green about this holiday? You'll be travelling large distances and getting to see a lot of the world mostly by wind power.

Who is it for? For a long voyage you need to have money, time and a real desire to learn all about working a tall ship.

What's a green way of getting there? You can arrange things to join ship at a British port.

How much pollution am I saving? Compared with the same distance by air you'll be saving something like four tonnes of carbon dioxide.

Anything not green about it? No.

What time of year? Any and all times of year.

What price? For a voyage around the world, or equivalent, think in terms of £20,000 or more.

Where can I get further information? www.picton-castle.com Tel. +1 (902) 634 9984; for information about other tall ships, events and races see www.sailtraininginternational.org

How do I make a booking? There's a form on the website.

Appendix 1

Your green travel calculator

Here's a calculator to tell you roughly how much pollution any journey will create. It's necessarily only a broad picture because your personal total will depend very much on variables such as the load factor of the plane, coach, train or car that you're travelling in and the route taken.

Nevertheless, the general message is pretty clear – go by coach, train or sailing ship whenever possible. If you want to go by car make sure (a) its fuel consumption is among the lowest possible, and (b) every seat is taken. Although a full medium-sized car produces less pollution per mile than a plane, don't forget that a car usually has to travel a greater distance to get to the same destination. And if there's only you in a large car then it's by far the most polluting form of land transport. For more information on cars see *Part One*. Avoid flying. Don't go on an ordinary liner as a way of getting to, say, North America; take a sailing ship instead. And never cruise at all except, once again, by sailing ship.

Destination	Carbon dioxide emissions for a one-way journey by:					
(capital city)	**Air**	**Car**	**Train**	**Coach**	**Cruise**	**Sailing**
	(Kg of CO_2 per passenger)					
Belgium/France	160	55	45	30	–	–
Ireland	240	80	65	45	–	–
Switzerland	380	80	90	60		
Germany	465	155	110	75		
Austria/Spain	625	210	150	100		
Italy/Poland	740	250	175	120		
Tunisia	925	310	250	180	1800	200
Greece	1200	400	290	200	2500	250
Egypt	1500	665	500	400	3000	300
UAE	2100	835	650	500	4000	400
USA/Canada	2300	–	–	–	3000	300
India	2600	900	850	550	5000	525
Jamaica	3000	–	–	–	4000	425
Mexico	3500	–	–	–	4500	475
Thailand/Sth Africa	3700	1250	1150	800	4000/ 7000	450/ 750
Australia	5400	2000	2000	1300	11000	1000

Notes:

1 These figures are calculated for travel from London or the south-east.

2 Cars. The figures are based on the average number of passengers in a medium-sized car which is 1.56. Of course, it's impossible to have 1.56 people in a car, but you get the broad idea. Obviously if there are three of you in the car you can more or less halve the figure. And if you have four people in a low-consumption car such as a Toyota Prius you can halve it again.

3 Boats. It has to be kept in mind that while planes fly fairly directly to their destinations other forms of transport have to take more convoluted and therefore longer routes. A cruise ship sailing from the UK to Greece, for example, will cover almost double the distance. The figures for sailing ships are somewhat notional since we don't know the size of the boat, how many people are aboard nor how often it will be necessary to run the engine when the wind is insufficient. But they give you an idea of the scale of difference between a cruise ship and a sailing boat.

4 General. These figures should only be treated as a rough guide. In many cases it isn't very practical to take, say, the train rather than the plane. And some overland routes would, of course, require one or more legs to be done by ferry, thus increasing the total amount of pollution.

Appendix 2

Green travel organisations

Here you'll find a whole range of organisations connected with green tourism, ranging from tour operators to campaign groups. All of them have something to contribute. But the fact that a company is included here doesn't mean that everything the company does is green. And, in fact, some of the organisations are in opposition to one another.

When booking a holiday you still need to be discriminating. For example, long-distance holidays by air just aren't green, no matter which company offers them, unless (and this is rare) the environmental benefits significantly outweigh the damage you'll do be doing when you travel. Ask yourself:

- What am I going to be doing when I arrive that's so beneficial it will more than make up for the pollution created?
- Is there an alternative nearer to home that would be better for the environment on balance?

Travel agents and tour operators

www.adventurecompany.co.uk Tel: 0845 450536. Specialist in small groups and with an interesting Europe content. Has launched a charitable foundation to fund projects in some destinations and claims to environmentally audit and offset all its trips.

www.audleytravel.com Tel: 01993 858000. Mostly long-haul tailor-made packages but has a stated inclination towards responsible tourism projects.

www.ecobookers.com Online green accommodation specialist.

www.forestholidays.co.uk Tel: 0845 130223. Forestry Commission tour operating arm. Offers camping and cabin holidays all over the UK.

www.mhrtravel.com Tel: 01704 57776. Has a long-established tree-planting offset scheme.

www.mighty-oak.co.uk Tel: 07890 698651. Isle of Wight operator offering tree-climbing packages with hammock accommodation.

www.naturetrek.co.uk; info@naturetrek.co.uk Tel: 01962 733051. Wildlife-watching holidays world wide.

www.organicholidays.co.uk Tel: 01943 871468. Directory.

www.responsibletravel.com Not a tour operator but a directory of holidays that promise to be more environmental, culturally sensitive, and beneficial to local people.

www.travel-quest.co.uk. Online only directory. Put 'green' or 'eco' holidays into its search facility

Other organisations

www.abta.com, The Association of British Travel Agents. Has founded, with others, the well-respected Travel Foundation (www.travelfoundation.org.uk) which aims to encourage sustainable tourism. ABTA is also a member of the Department of Trade and Industry's Sector Sustainability Pioneer Group.

www.aito.co.uk The Association of Independent Tour Operators. Has launched a sustainable tourism scheme.

www.allorganiclinks.com Includes some travel in its varied listings.

www.atkinsglobal.com Environmental risk assessment specialists for the travel and tourism industry.

www.bornfreefoundation.org.uk An international charity campaigning on behalf of animals in captivity.

www.carshare.com Tel: 08700 780225. Directory of UK and Ireland car sharing sites, especially for large businesses and organisations.

http://ec.europa.eu/enterprise/services/tourism/tourism_sustainability_group.htm. Environmental experts working together to draw up Europe's blueprint for the sustainable future of tourism. It's aims are to:

- Ensure the long-term competitiveness, viability and prosperity of tourism enterprises and destinations
- Provide quality employment opportunities offering fair pay and conditions and avoiding all forms of discrimination
- Enhance the quality of life of local communities through tourism and engage them in its planning and management
- Provide a safe, satisfying and fulfilling experience for visitors without discrimination
- Minimise pollution and degradation of the global and local environment
- Maintain and strengthen cultural richness and biodiversity and contribute to their appreciation and conservation.

www.ecotravelling.co.uk Online ecotravel articles, advice and comment.

www.eia-international.org Tel: 020 7354 7960 The Environmental Investigation Agency. Goes undercover to expose environmental crime. You can support the work of the agency by booking certain holidays with Naturetrek (see above).

www.english-nature.org.uk

www.evansjones.co.uk Helps organisations move away from car transport by developing cycling and car sharing schemes.

www.greentraveller.co.uk Award-winning online eco-magazine.

www.nationaltrust.org.uk Tel: 0870 4584000

www.nts.org.uk Tel: 0844 4932100 (National Trust Scotland)

www.liftsharesolutions.com Promotes car-share projects.

www.resurgence.org Online and print magazine dedicated to all things eco.

www.sascotland.org Tel: 01316 662474. Scotland's organic Big Daddy.

www.seat61.com An extremely well-researched website for national and international train travel.

www.soilassociation.org.uk Tel: 01173 145000 Big Daddy of UK organics.

www.travelwise.org.uk Grouping of local authorities and other large bodies working together to promote sustainable travel.

www.uk-energy-saving.com Useful site covering all aspects of eco-life.

www.vso.org.uk Volunteer organisation with an emphasis on responsible travel (see also www.beso.org).

www.wildlifetrust.org Tel: 00 1 2123804460. USA-based international wildlife protection organisation.

www.wwoof.org.uk Working volunteers on organic farms.

www.wwt.org.uk Tel: 01453 891900 The Wildfowl and Wetlands Trust based at the famous Slimbridge Wildfowl Centre.

Tourism for Tomorrow Awards

Every industry nowadays has its awards and tourism is no exception. These awards were introduced by the World Travel and Tourism Council (WTTC).

These were the most recent winners at the time of writing this book.

THE DESTINATION AWARD

Won by the Great Barrier Reef Marine Park, Australia

The Great Barrier Reef is actually the largest World Heritage site on the planet and one-third of it is now protected. The Marine Park won not simply because it's a wonderful place, but also for the fine balance that's been struck between the environment and visitors (who, after all, generate vital income for conservation).

www.gbrmpa.gov.au

THE CONSERVATION AWARD

Won by the Aspen Skiing Company, USA

Since 1997 the Aspen Skiing Company has launched various initiatives to address the concerns environmentalists have about ski resorts. They include a wind-powered lift as well as energy generation from hydro-electric power and sunshine.

www.aspensnowmass.com/environment

THE GLOBAL TOURISM BUSINESS AWARD

Won by Lindblad Expeditions

In 1958, Lars-Eric Lindblad pioneered a style of travel we now call eco-tourism. Today, his son, Olof Lindblad runs a travel company, Lindblad Expeditions, dedicated to continuing his father's vision. In the Galapagos, for example, the company set up the Galapagos Conservation Fund and invited clients to donate. In 10 years the fund has raised almost £2 million. Other destinations visited by Lindblad Expeditions include Antarctica, Baja California, Alaska and the Arctic.

www.expeditions.com

THE INVESTOR IN PEOPLE AWARD

Won by Nihiwatu Resort, India

Nihiwatu is a luxury resort that has been named by Tatler Travel Guide as one of the '101 Best Hotels In the World'. In 2006 it was also top of Condé Nast Traveller's 'Green List'. Yet it comprises only 14 villas, mainly constructed of thatch, wood and bamboo. It won because of the contribution the resort makes to the wellbeing of the local community.
www.nihiwatu.com

If you'd like to see all the past winners of the Tourism for Tomorrow Awards go to www.tourismfortomorrow.com

THE VIRGIN HOLIDAYS RESPONSIBLE TOURISM AWARDS

The Virgin Holidays Responsible Tourism Awards are a collaboration between the online travel agent responsibletravel.com, the Times newspaper, the Geographical Magazine and the World Travel Market. The guiding principle is that all types of tourism should respect and benefit destinations and local communities.

These were the most recent winners at the time of writing:

Overall winner: The New Forest www.thenewforest.co.uk

Best destination: The New Forest

Person who has made the greatest contribution to responsible tourism: Anthony Climpson, The New Forest.

Best tour operator: Gecko's Adventures, Australia www.geckosadventures.com

Best large hotel: Apex Hotels www.apexhotels.co.uk

Best small hotel: Finca Esperanza Verde Ecolodge, Nicaragua www.fincaesperanzaverde.org

Best low carbon transport and technology: Eurostar, UK www.eurostar.com

Best in a mountain environment: Explorandes, Peru www.explorandes.com

Best in a marine environment: blue o two www.blueotwo.com

Best for poverty reduction: Borana, Kenya www.borana.co.ke

Best in a park or protected area: La Ruta Moskitia, Honduras www.larutamoskitia.com

Best for conservation of an endangered species: Grootbos Nature Reserve, South Africa www.grootbos.com

Best for conservation of cultural heritage: Andaman Discoveries, Thailand www.andamandiscoveries.com

Best volunteering organization: Azafady, UK/Madagascar www.azafady.org

Index

Hands-On Holidays

Looking for a life-changing experience? Just a few weeks to spare?

Whether you want to make a difference, or just do something different, Hands-On Holidays is the definitive guide to finding the most rewarding short-term breaks worldwide.

Catering for all ages, budgets and interests, hands-on holidays are for responsible travellers who want to use their holiday time to do something positive. They offer a real opportunity to get involved, to learn and contribute, and to enjoy a more authentic travel experience. From wildlife research and community schemes, to charity challenges and volunteering at music festivals, Hands-On Holidays is packed with holiday ideas that you won't find anywhere else.

Hands-on holidays last from just a few days to a month, so are perfect even for those with commitments at home, professionals on annual leave, and students during their vacation.

Author: Guy Hobbs
Published: July 2007
ISBN: 978 1 85458 373 4

The Scottish Islands

The best-selling and most comprehensive guides available to Orkney and Shetland, and Skye and the Western Isles.

These islands present visitors with a rich variety of terrain, an abundant archaeological legacy, stunning coastal scenery and unique flora and fauna, guaranteeing a spectacular and magical experience.

The Scottish Island guides are the ideal companion for a trip to the islands, packed with comprehensive information on the most scenic walks, the liveliest festivals and ceilidhs, the best places to see wildlife, the most impressive castles and monuments, and the best fishing, sailing and diving locations.

Author: James and Deborah Penrith
Published: July 2007
ISBN: 978 1 85458 371 0

Author: James and Deborah Penrith
Published: July 2007
ISBN: 978 1 85458 370 3

www.crimsonpublishing.co.uk